Fly Tying

Made Easy For Beginners

Dedication

For Uncle Dale and Aunt Juanita Baumgartner
for being wonderful role models, for inspiration,
and for the fabulous laughter over the years.

Books By
Randall Kaufmann

American Nymph Fly Tying Manual
Lake Fishing With A Fly
The Fly Tyers Nymph Manual
Tying Dry Flies
Bonefishing With A Fly
Tying Nymphs
Tying Dry Flies, Revised Edition
Fly Patterns Of Umpqua Feather Merchants
Fly Patterns Of Umpqua Feather Merchants, Second Edition
Bonefishing!
Tying Nymphs, Second Edition
Tying Dry Flies, Third Edition
Tying Nymphs, Third Edition

Fly Tying
Made Easy For Beginners

Randall Kaufmann

Foreword - Randy Stetzer

Western Fisherman's Press

2002

Acknowledgments

The following family and friends have contributed to this project, helping to make it possible: Jay Daley, Jim Black, Kevin Erickson, Barry Garriety, Hal Janssen, Joel La Follette, Steve Lent, Tiesha Kaufmann, Tresa Kaufmann, Ken Mitchell, Jerry Swanson, the innovative tyers whose patterns are featured, and you, the reader, for your enthusiasm and ultimately making it all possible.

Special thanks to my brother, Lance, for picking up all the pieces; my wife, Mary, for not losing sight of me; my mother, Oda, for inspiration; Bob Rector, for editing and proofing; Joyce Sherman, for production and managing the project; Randy Stetzer, for writing the foreword; and Mike Stidham, for art work. Thank you very much.

All photos and flies by Randall Kaufmann unless otherwise noted.

Published by
Western Fisherman's Press™
P.O. Box 357
Moose, Wyoming 83012
www.fishermanspress.com

Distributed by:
Kaufmann's® Streamborn,® Inc.
PO Box 23032
Portland, Oregon 97281-3032
(503) 639-6400 phone
(503) 684-7025 fax
e-mail: books@kman.com
www.kman.com

Printed in Thailand
10 9 8 7 6 5 4 3 2 1

Library of Congress Catalog Card Number: 2002-190403

International Standard Book Number: 1-885212-19-4

Contents

Foreword

Getting started tying flies can be overwhelming, even daunting to the uninitiated. Just the basics of selecting equipment and materials can discourage all but the most tenacious. The rewards of perseverance, however, are many. There are several reasons to tie your own flies, from saving money, to making money, or simply to add to the overall experience. Many of the most successful anglers tie their own flies because they cannot always buy exactly what they want.

You are less likely to become frustrated or discouraged and more likely to have fun and find success, if you begin with solid fundamentals. Like any endeavor, the most expedient and efficient path is to find a mentor to show you the ins and outs of the game.

There is no better mentor than Randall Kaufmann. Randall began tying as a hobby at 14, became professional at 15, searched out the world's best materials, tyers and trout, and at 22, with his brother, Lance, and their mother, Oda, began Kaufmann's Streamborn, Inc., a worldwide fly fishing supply house.

Randall has experienced and excelled at all levels of fly tying, from tying thousands of dozens commercially to creating some of the most effective patterns known, such as the Stimulator and Kaufmann Stone. Commercial tying taught Randall how to streamline his efforts and develop efficient methods for tying well-proportioned, durable flies *quickly.* Teaching fly tying to hundreds of students taught him how to best present his techniques to others. It is rare to find someone who can tie a wide gamut of patterns well, and be able to convey such knowledge to others in a clear, logical, and easy-to-understand format.

The first three chapters on tools, hooks, and materials alone make this book worthwhile, but it is Randall's tying technique and presentation with words, light and lens that elevates this instruction book far above all others. As you turn the pages, the presentation and tying sequence are exactly like one would use to tie a fly from start to finish. From setting up the bench, to mounting the hook, securing the thread, and tying materials onto the hook, to completing the many flies, the sequence is perfectly logical and focused.

Fly Tying Made Easy For Beginners is like tying flies with Randall Kaufmann—your own personal instructor—explaining techniques and leading you through the tying process of 28 popular flies. It doesn't get much better! When you consider Randall's unmatched experience, expertise, and genuine love for what he does, both beginning and expert tyers benefit immensely from having him as a mentor. Follow Randall's no-nonsense approach and concise step-by-step instructions, and you should be tying great flies in short order.

Randy Stetzer
Portland, Oregon
March 2002

Brian O'Keefe

The flies…creating and organizing your collection allows endless pleasure.

The fish…like this rainbow trout, are all special. Please use barbless hooks and handle them gently and as little as possible.

Introduction

Fly tying is an intrinsic component of fly fishing that can be enjoyed at any time without leaving the house. Tools and materials are portable, and you can take them anywhere you go. Over the years, to the amazement of many, I have set up the vise on airplanes, or replaced lost flies in steelhead camp by candle light, or matched-the-hatch streamside by lantern light. Tying flies always gathers an admiring crowd and fosters camaraderie and creativity.

Unlike fly fishing, which requires water and flies, fly tying does not require fishing. Some tyers seldom tie a fly to a leader. Some carry the hobby to an art form. Others tie out of necessity. The lucky ones tie for both pleasure and necessity, which means they're having big time fun! Fly tying often becomes its own mania. It is truly a unique escape from the ordinary.

Like anything, fly tying is easy if you know how. Locating the right source of knowledge and instruction can be as daunting as acquiring the skills you seek. I was lucky to begin tying at age 14 and selling flies to local sporting goods stores by age 15, but I didn't really know how to *produce* flies. When I was 16, I purchased a car with money saved from selling flies. Summers were spent tying flies and fishing the West. This allowed me to become acquainted with, and watch, many of the great fly tyer-anglers of the time practicing their craft. Like a fingerprint, every tyer has his or her own style and technique, and often specializes in a particular style of fly, whether it be nymphs, dries, streamers, steelhead, saltwater, etc. I adapted the best techniques from everyone, and that is what is condensed into this manual. Remember, no one has all the answers to any one subject. I urge you to read all you can, watch others tie, and think outside the ring of the rise.

The basic information and tying techniques needed to begin actual fly construction are explained in chapters 1 through 5. Practice these techniques *before* progressing onto the demonstration flies. Like anything new, your hand-mind coordination may not be perfect, but it gets better with practice.

We begin actual fly construction with simple, yet effective flies and progress to more complex ones. A book this size cannot cover everything and beginning tyers should concentrate on success. These pages do, however, *build a strong professional foundation to expand upon.* After all 28 demonstration flies have been mastered you may wish to continue with my other tying books, *Tying Nymphs* and *Tying Dry Flies*. These build upon techniques explained here and detail more intricate patterns, including the Kaufmann Stone, Scud, Floating Dragon, Humpy, Wulff, Irresistible, Stimulator, Paradrakes, Dave's Hopper, etc.

The goal of this book is to make beginning tyers successful at tying their *first* fly, and to show all tyers professional tricks that help make tying fun and easy. If you are successful, you are having fun, and you may discover some unique insights of your own.

Randall Kaufmann,
Moose, Wyoming
February 2002

Chapter 1

Tools

Tools make the difference between success and failure, fun and frustration. Like carpenters, fly tyers need good tools. A vise that holds the hook tightly without slipping, fine-point scissors that cut tiny fibers accurately, hackle pliers that hold hackle securely, and other accessories allow you to concentrate on and complete your work.

Application Jar

This is an effective way to dispense and keep lacquer from evaporating and scenting the air. Attach the jar to the shaft of your vise with rubber bands or tape, or secure it in a Renzetti Tool Caddy, and it is always handy.

Blender

This is useful for blending colors, textures, and sparkle together. Any texture or color is instantly available to the fanatic tyer. Natural dubbing materials (goat, muskrat, HareTron) blend much more easily than synthetics. Long strand synthetics should be cut into lengths of one inch or less before blending.

Bobbin

A good bobbin is a must. Its purpose is to hold the thread spool, to release thread as you need it, and to keep constant tension on the thread. Operating techniques are detailed in the Basic Tying Techniques chapter. I believe the Matarelli is the best stainless steel bobbin. Tiemco offers a superb ceramic bobbin that is nearly indestructible. Both are lightweight, durable, finely polished, and easily adjustable.

Bodkin/Half Hitch

This is a needle point used for picking out dubbing, lacquering heads, freeing up hackle, and, if it has a hole in the end, for tying half hitches. Select one with a hexagonal shaft so it does not roll off the table.

Cements

Lacquer is applied to secure the thread knot at the head just behind the hook eye. It should be fast drying and of relatively high gloss. There are many such products available. Vinyl cement (Flexament) is used for reinforcing or stiffening feather material such as turkey wing quills. Apply vinyl cement with a small brush or a bodkin. Epoxy puts an impenetrable gloss on thread heads, but it must be applied thinly. When applied to wingcases, it creates a shiny, distended pre-hatch appearance. Epoxy is not usually applied to small heads. Lacquer is sufficient. Zap-A-Gap secures eyes and wire.

Application jar keeps a tight lid on and dispenses lacquer.

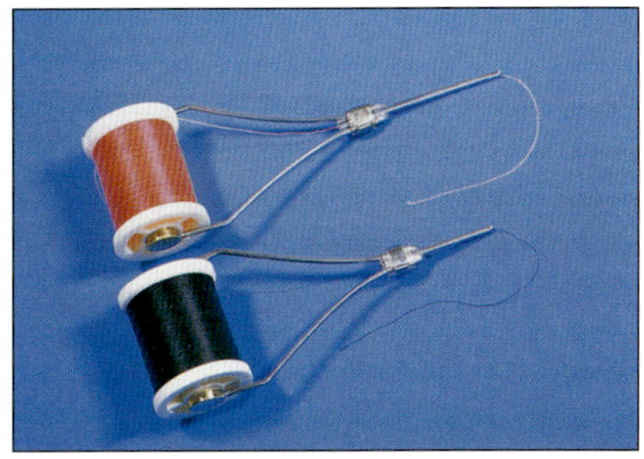

Use a short shaft bobbin for small flies; a long shaft for long-shank flies.

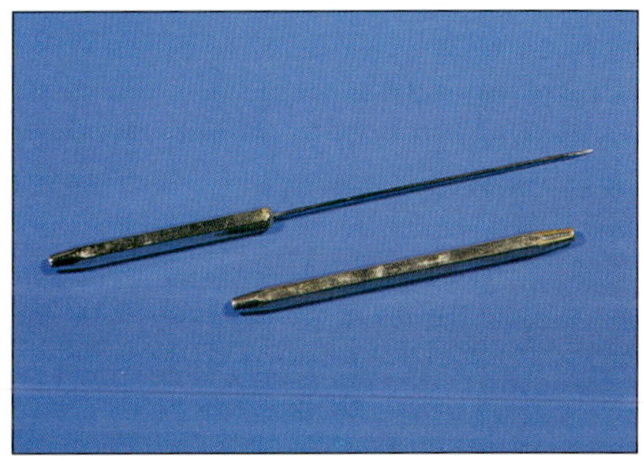

Tyers use a bodkin (top) on every fly to apply lacquer to the head. Half hitch tool (bottom).

Take a turn of thread around a half-hitch tool and slide half hitch off end.

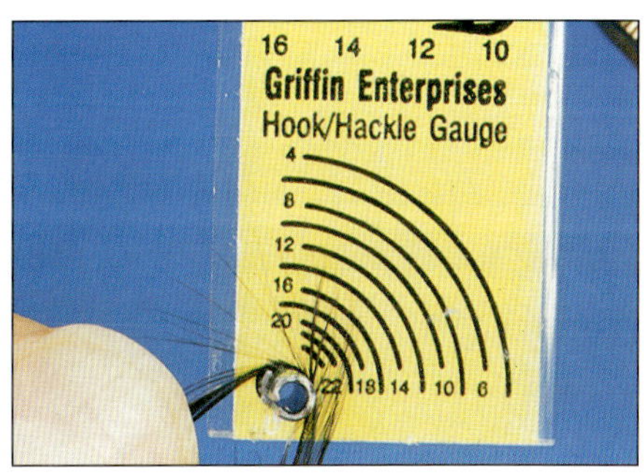

Hook and Hackle gauge is a useful learning aid.

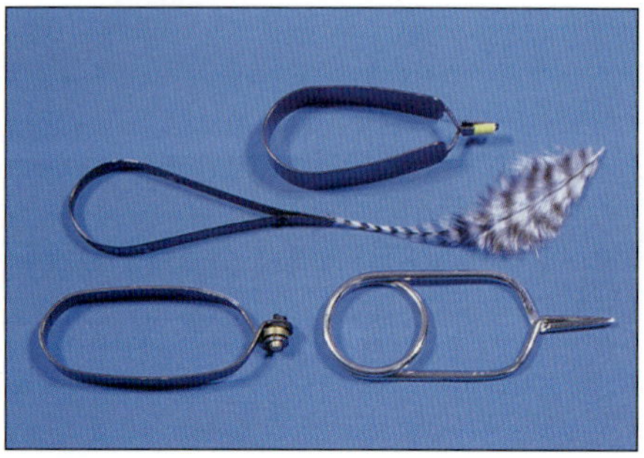

Hackle pliers are able to grip the shortest and smallest items.

Hook and Hackle Gauge

Compact and efficient, the Hook and Hackle Gauge attaches to the vise shaft and allows you to select size 4 to 24 hackles quickly and accurately. It also measures hook size. This is a *general* guide and learning aid. Eventually you can work without it.

Hackle Pliers

Hackle pliers are pinchers used to hold onto material while you wrap it around the hook. They are especially useful for "short" ends that are difficult to hold onto with fingers. Most have a loop that your finger fits into, which helps you turn them. When pliers can be of help, use them. There are several styles.

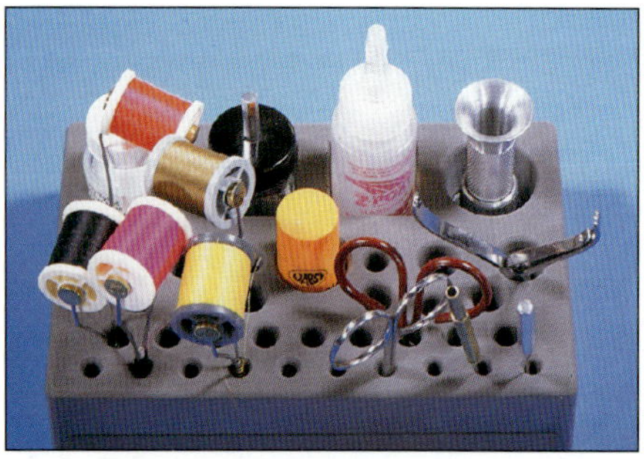

Tool caddy organizes tools so they are always convenient. Nice to have!

Light

If you can't see clearly, you can't tie. If possible, get a light that has a concentrated beam of light, burns clear, cool, and bright and has an adjustable arm. If one light isn't enough, get another—remember, this is supposed to be *fun!* Some lights offer built-in magnifiers.

Mageyes

Seeing clearly makes fly tying possible. If you have trouble, a hands-free magnifier that slips onto your head, attaches to the vise, or stands alone may be ideal. Other aids are also available.

Adjustable finger hole scissors with a sharp, fine point.

Material Clip

There are two types: spring and the Thompson slot style. This sanity saver attaches to the vise and holds tinsels and other materials out of the way until you are ready for them.

Tool Caddy

The perfect organizer for your tying bench. Keeps tools secure, organized, and handy and keeps cements from spilling.

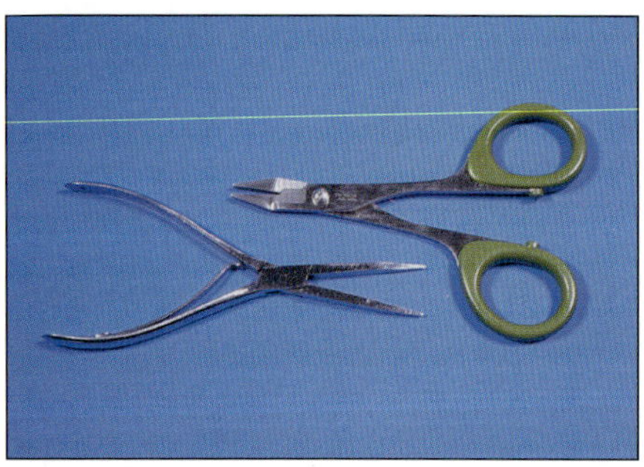

Smooth-nosed pliers belong on the tying bench and in your vest.

Scissors

Scissors should have a *fine* point, thin blade, and finger holes large enough so you can carry them on your hand. When you carry scissors on your hand, you always have them ready to cut without wasting time looking for them, picking them up, and putting them down. Scissors can be carried on your thumb and first finger or on your thumb and any other finger you feel comfortable with. You can also carry them on your thumb and in your palm, inserting a finger when you need to make a cut. All cuts should be made with the tip of your scissors. If possible, make only one exact cut. When making close cuts, steady the scissors against the hook or vise while holding onto the item to be cut. Cuts can only be as close as the blades are thin. Cutting paper or wire with precision scissors is a good way to ruin them. Use another pair for cutting wire.

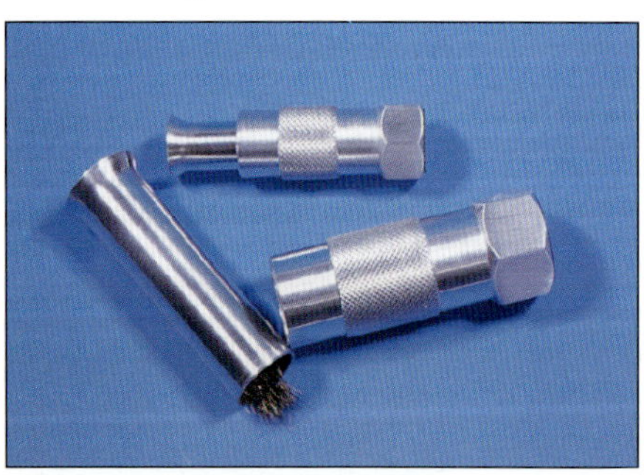

Stacker, a pull-apart cylinder that "stacks" or evens hair.

Smooth-Nosed Pliers

Smooth-nosed pliers with a fine point are used for smashing barbs *before* you tie the fly. They are also used to flatten lead bodies. Tiemco DeBarb pliers are strong and the point is blunt so they are less dangerous.

Stacker

A stacker is indispensable for evening the tips of hair tails and wings, especially when tying dry and steelhead flies. Larger-diameter stackers allow hairs to spread apart and "fall" or even up more quickly.

Threader-Cleaner

This is a simple tool used to unclog wax that builds up and to pull thread through the bobbin shaft. Thread the bobbin by inserting the end of the threader-cleaner with the wire loop into the bobbin. Insert the thread through the wire loop. Pull the wire loop and the thread through the bobbin shaft.

Vise

A vise should hold a wide range of hooks *securely.* The Regal, Dyna-King Barracuda, Renzetti Master's, and many other top-of-the-line vises hold a hook so tightly that you can break the hook before it slips from the vise jaws. The Regal holds the largest and smallest hooks without adjustment. The Dyna-King requires minimal adjustment. I like base vises because you can set them up anywhere. The Regal bronze, Dyna-King Barracuda, and Renzetti have heavy bases that do not rock or wobble. All are available with a C-clamp, which is more secure and offers greater height adjustment. The Thompson A is the old standby and the best inexpensive vise. I do not recommend foreign vises. Besides the quality issue, you cannot always get parts. Adjust the height of your vise so there is eight to 10 inches of clearance between the table's surface and the jaws.

Trim Bag

This is a wire hoop basket that attaches to the vise. It acts as a waste container and catches dropped tools and materials and spilled lacquer.

Wax

Wax can ease dubbing application and increase thread strength. Tyers should be using pre-waxed thread when possible. Sometimes additional wax makes it easier to dub coarse materials.

Most Important Tools

The following tools should be on every tying bench: vise, bobbin, threader-cleaner, scissors, stacker, hackle pliers, smooth-nosed pliers, bodkin, and material clip.

Fly Tying Storage Cases

There are many wooden boxes and soft material cases available for storing and organizing supplies. Some are designed for travel, others for home use. Organization is a key factor. If you can view and access your inventory quickly, you spend more time tying.

Setting Up the Tying Bench

Most tyers begin tying at the kitchen table or office desk. It usually does not take long to outgrow or to be evicted from such quarters. A desk devoted solely to tying is ideal, but a card table can suffice. If you keep your vise set up permanently, you can take advantage of short periods of free time. If you have to set up and tear down your tying area every time you want to tie a couple of flies, you may tie infrequently. You may find that a few minutes at the vise is a great stress reducer and a way to vicariously escape to your favorite fishing waters. If you just tie a fly or two here and there, you may be surprised how many you accumulate in a short time.

A light background allows you to see hooks, tools, and materials clearly. Light blue poster board is perfect. Be certain to keep all dangerous objects and materials away from children and pets.

You should be able to break a hook before it slips in the vise.

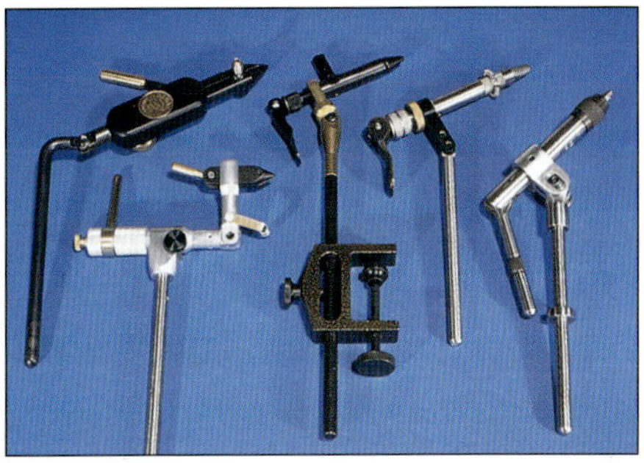

A vise is the most important tool. It should hold a hook securely.

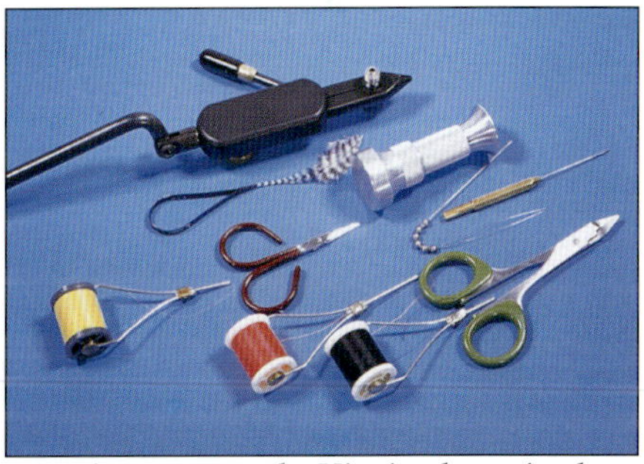

Most important tools. Vise is a base vise; base is not shown.

Storage cases organize and carry enough material to tie hundreds of flies. JW Pro shown.

Organize and leave your tying bench set up so it is always available. Note trim bag.

Chapter 2

Hooks

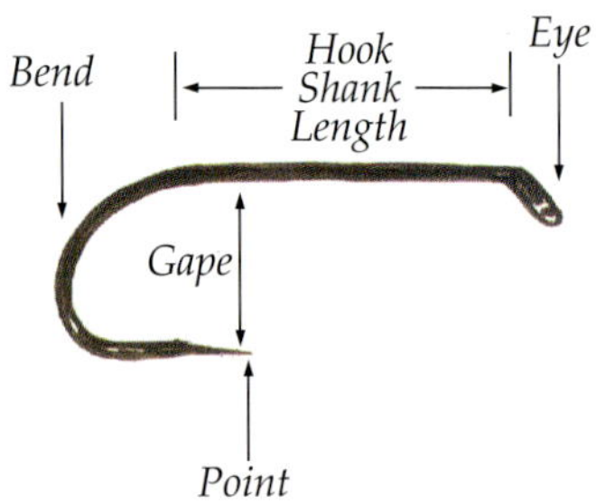

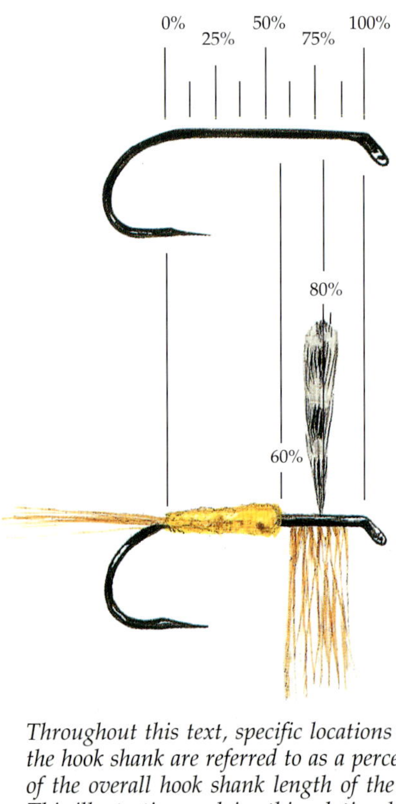

Throughout this text, specific locations along the hook shank are referred to as a percentage of the overall hook shank length of the hook. This illustration explains this relationship.

There are several premium hook manufacturers, including Partridge in England and Dai-Riki, Daiichi, and Tiemco in Japan. These manufacturers use the latest technology, hook designs, and the strongest tempered steel. The hook points are sharp, the eyes are closed tightly without any rough edges, and the finish is blemish free. Fishing the best possible hook is another way to maximize your angling success.

The major criteria for selecting hooks are size (determined by gape), length, wire diameter and shape. All other considerations are minor, but an understanding of terminology helps to understand hooks and labeling.

Gape

The gape is the distance between the hook point and shank. For the most part, manufacturers adhere to a standard gape for each size of hook, making the gape the tyer's standard unit of measure. Hooks are sized by numbers, 3/0, 2/0, 1/0, 1, 2, 4...12, 14, 16, 18, 20...to 32, which is the smallest. Odd-numbered hooks smaller than size 3 are seldom manufactured. As hooks get smaller, the difference between sizes is reduced—that is, the difference between sizes 10 and 12 is greater than the difference between sizes 20 and 22. Because there is a substantial difference in size on larger hooks, they are available in odd numbers. There are larger sizes than 3/0 available.

The larger the number, the smaller the gape. A size 8 is larger than a size 10. The hook length, wire diameter, bend, and eye design may vary, but the gape is *mostly* uniform for any given hook size. A standard hook gape is roughly half the length of a standard hook shank.

If a hook gape is too narrow, hooking fish may be difficult. Tyers should remember not to impede the hook gape with bulky materials, which can reduce hooking capability. If the gape is too wide and the hook rides inverted (hook point up), it can penetrate a fish's eye or brain, causing serious injury or death. Wide gape hooks should be avoided.

Wire Diameter

Manufacturers have standardized wire gauges *somewhat;* a size 12 standard fine wire dry fly hook *usually* has the same diameter wire regardless of its manufacturer or configuration. Heavier gauge wire is used for subsurface flies; it may be chosen for strength, especially for saltwater use. An X system is used to denote wire diameter that is smaller or larger than standard gauge. Each X denotes one size, even though hooks are seldom manufactured in small odd sizes. A 1X fine size 14 hook has the same diameter as a hook one time (1X) smaller, or a size 15. A 2X fine size 14 has the diameter of a standard size 16. A 2X heavy size 12 is the same diameter as a size 10. Any fly may be tied on hooks that are stronger, heavier, lighter, shorter, or longer than standard.

Shank

The shank is measured from the eye to the beginning of the bend. There is also an X labeling system for hook shank length. A 2X long size 10 means the hook length is the standard length of a hook two sizes larger, a size 8. A 6X long size 10 denotes the shank length of a size 4 standard length hook. Depending on the actual length of the natural being imitated, dry flies may be tied on 1X, 2X, or 3X long hooks, nymphs on 1X to 6X long hooks. Streamers, leeches, etc. are usually tied on 3X, 4X or 6X long hooks. Steelhead hooks are 1X or 2X long, have an upturned loop eye, and are usually black in color.

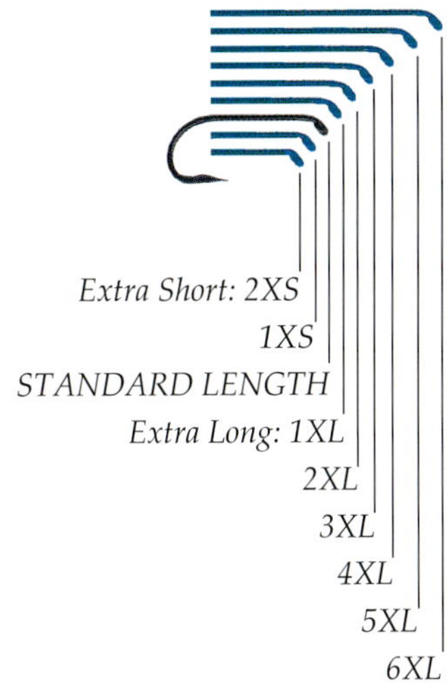

Hook shown is an actual size 10 Tiemco 900BL, standard length hook. Shanks and eyes are shown in blue for shorter and longer hooks.

Eye

The most common eye designs are turned down (TD), followed by turned up (TU) and straight. These may incorporate a tapered, loop, or ball design. Standard dry fly hooks usually have a turned-down tapered eye (TDTE). Turned-up eyes are usually preferred for steelhead and salmon flies. The shape of the eye, in conjunction with your leader knot, helps determine how the fly rides on or in the water.

I prefer to attach turned-down and turned-up eye hooks with a Turle knot. The Turle knot allows for a straight pull (it is an extension of the hook shank) and does not rotate or swing around the eye like a clinch knot. If a fly is cocked at an angle, it may not fish properly, or the leader may become wedged in the corner of a flawed hook and break easily. Because the Turle knot secures itself behind the hook eye, it does not create a knot in front of the fly. Depending on leader diameter, a clinch knot tied on a size 20 hook can enlarge the imitation by over 25 percent and alter the surface impression. Eyes that are not completely closed or that have a rough closure can cut thread and make tying a neat head difficult. Check hook eyes; they are important.

Point

The point is the most fragile part of a hook, and care should be taken to avoid damaging it, especially in the vise and on back casts. Larger hooks and heavier flies can also damage rods when they hit during windy conditions or power casting.

During tying, hook points should not be buried in the vise jaws. Chemically sharpened hook points do not need sharpening before you fish them, but they should be checked often for broken or dulled points while you are fishing. A traditional method for determining if a hook point is sharp is to drag the point across your thumbnail. If the hook leaves a scratch, it is fairly sharp. If it stops dragging, it is very sharp.

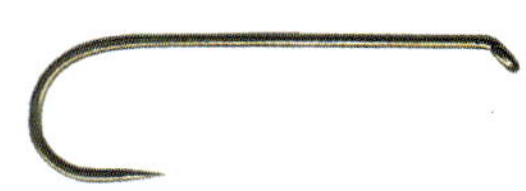

Turned-down tapered eyes are the most common eye design. Tiemco 5263BL is shown.

Barbless Hooks

There are only positive reasons to use barbless hooks. A barbless hook penetrates a fish's mouth more easily than a barbed hook, which means more fish are hooked and few, if any, come unhooked. Fish hooked with a barbless hook are easy to release; simply back the hook out in the same direction it went in. This can usually be accomplished instantly and with one hand while keeping the fish in the water without handling or squeezing it, which means fish experience less stress and have a better survival rate. It also means you spend less time reviving fish and more time fishing.

Barbless hooks are also easy to remove from yourself! When you hook yourself, and you will, you'll be doubly glad it's barbless! If the hook is not barbless, smash the barb, using a smooth-nosed pliers, before you tie the fly. Better yet, buy them barbless.

Straight or ring eyes are sometimes used by tyers who are concerned about a turned-down eye blocking the hook gape and, consequently, the hooking ability of the hook. This can be a concern with some hook styles, especially size 18 and smaller. Tiemco 200RBL is shown.

Turned-up loop eyes are the accepted standard for steelhead, Pacific salmon, and Atlantic salmon flies. Tiemco 7999 is shown. Smash the barb.

Comparison-Substitution Guide

Tiemco	Mustad	Dai-Riki	Daiichi
900BL*	—	—	—
100BL*	94845*	320	1180
101	94859	310	1110
200RBL*	80050	270	1270
2457	80200	135	1150
2487BL*	—	—	1130
3761	3906B	060	1560
5212	94831	—	1280
5262	9671	730	1710
5263BL*	9672	710	1720
300	—	—	2430
7999	36890	899	2161

*Barbless

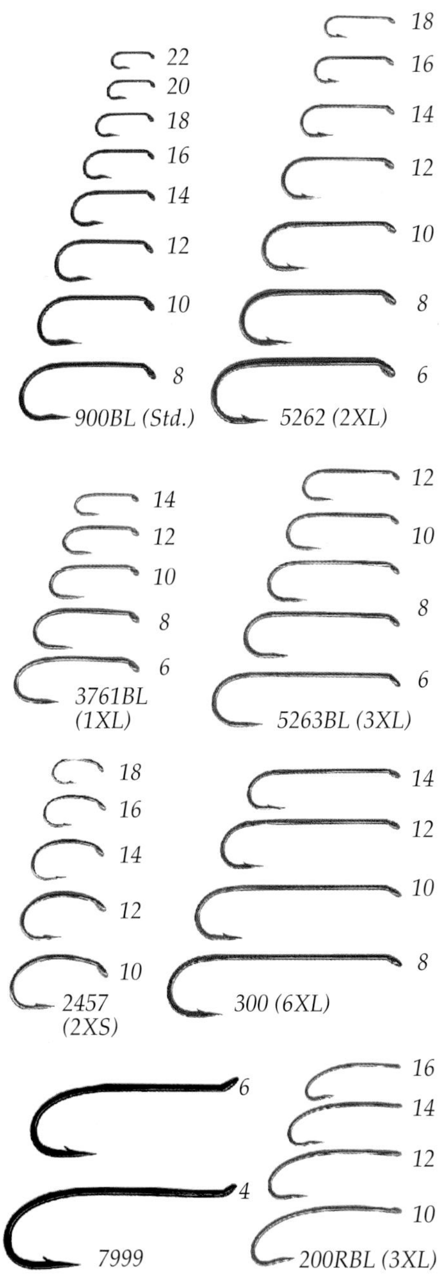

Recommended Hooks

My approach to hooks is the same as it is for fly patterns: keep it simple. I don't need nine different dry fly hooks any more than I need nine different damselfly nymphs. As a beginner, when I saw a pattern that recommended a specific hook, I thought I had to have it. The result? Wasted time trying to locate them, wasted money once I found them, and wasted hooks when I realized I didn't need or want them. The solution? Tie flies on *your* favorite hook, not everyone else's!

Pick one dry fly hook, which suffices for 90 percent of all dry flies. Pick a second, longer shank dry fly hook for stoneflies, hoppers, etc., and you cover 95 percent of all dry flies. Most nymph and streamer style flies combined require about six styles. We use seven different hooks for the 28 demonstration flies in this book. Similar instructions might call for 28 different hooks from five manufacturers! Collect all the hooks from all the manufacturers you like, but if you want to keep it simple, select one manufacturer.

I use Tiemco hooks. Tiemco pioneered the premium hook market and offers the largest selection of barbless hooks in contemporary styles and sizes, and they are easy to obtain. They may cost a bit more, but they are worth it. All hook numbers in this book are Tiemco.

Tiemco 900BL, sizes 8-20, *extra light wire (1X), standard length,* black, barbless, perfect bend (half circle), turned-down tapered eye (TDTE). This is my dry fly hook of choice, and it is occasionally used for nymphs.

Tiemco 100BL, sizes 8-24, *extra light wire (1X), standard length,* bronze, turned-down eye. An excellent all-purpose dry fly hook. If you have the 900BL, you do not need these.

Tiemco 101, sizes 20-26, *extra light wire (1X), standard length,* bronze, straight (ring) eye. Useful for tiny flies where the open ring eye allows a bit more hooking power.

Tiemco 200RBL, sizes 6-18, *3X long,* standard wire, bronzed, semi-dropped point, slightly humped shank, straight eye. A favorite for nymph and dry flies (adult stoneflies). Ideal when a *slightly* curved nymph is desired: caddisflies, stoneflies, mayflies, and leeches.

Tiemco 2457, sizes 10-18, *2X short, 2X heavy, 2X wide,* bronzed, curved shank, TDTE. Good choice for some scud and caddis nymphs when a curved body is desired.

Tiemco 2487BL, sizes 12-20, *fine wire, 2X short, 2X wide,* bronzed, curved shank, TDTE. Same as 2457 only fine wire and barbless. Pick one or the other, or use this hook for size 20 which is not available in the 2457.

Tiemco 3761BL, sizes 10-18, *1X long, 2X heavy,* bronze, sproat bend, TDTE. Good for general purpose nymphs (Pheasant Tail, Hare's Ear, Soft Hackle, etc.).

Tiemco 5212, sizes 6-12, *extra fine wire (1X), 2X long,* bronze, down eye. First choice for hoppers and adult drake patterns.

Tiemco 5262, sizes 8-18, *2X long, 2X heavy,* bronze, perfect bend, TDTE. Same as 5263 only slightly shorter at 2X long. When you need a hook shorter than 5263 and a bit more bite than the 3761BL, this is it.

Tiemco 5263BL, sizes 6-14, *3X long, 2X heavy,* bronze, perfect bend, TDTE. Identical to 5262, but with a longer shank. Many uses: stoneflies, dragonflies, leeches, streamers, and adult stones.

Tiemco 300, sizes 6-10, *6X long, 1X heavy wire,* bronzed, TDTE. For extra long nymphs (stoneflies), leeches, and streamers.

Tiemco 7999, sizes 2-6, *2X long, 1X heavy,* black, turned up tapered loop eye. For steelhead and salmon.

Chapter 3

Materials

Size 6/0 pre-waxed Danville is all-purpose thread. You should have several colors.

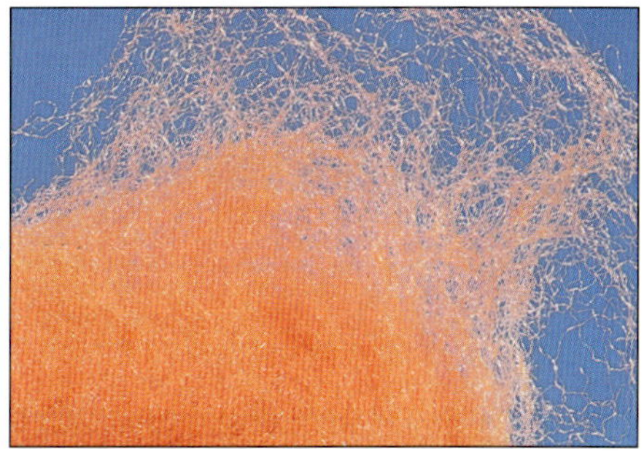

Dubbing material.

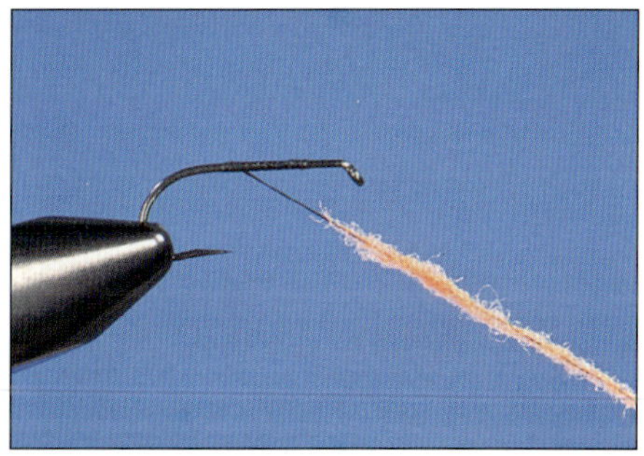

Dubbing material after it has been "dubbed" onto thread.

The dubbing strand above has been wrapped around the hook, creating a dubbed body.

Dubbed nymph, with natural dubbing, and dry fly abdomen, with synthetic dubbing.

Flies are only as good as the materials and workmanship that go into them. Select the best materials possible. Nymphs (subsurface imitations) and streamers (fish imitations) should be constructed with materials that enhance animation and visual realism. Scraggly nymphs are sometimes desirable. Dry flies (surface imitations) are often designed to imitate specific and delicate insects that fish inspect close up. This is especially true of trout in slow, clear water. Stiff hackle, water-repellent materials, light wire hooks, and floatants help keep dry flies afloat.

Open any catalog specializing in fly tying supplies. The selection is daunting. It is beyond the scope of this book to detail everything, but many of the most-often-used materials are described, along with what they are used for. If you collect all the materials and the various colors described here, you would be able to construct hundreds of patterns. The materials list at the end of this chapter includes everything needed to tie almost every fly pictured in this book.

Thread

Most flies require fine-diameter thread that does not build up bulk. Size 6/0 waxed Danville (Flymaster) is the accepted standard, but size 8/0 is useful for flies size 18 and smaller.

Some hair flies require stronger thread, such as size 3/0 or Kevlar. Spun and trimmed deer and caribou can be tied with 6/0, but some tyers prefer Kevlar, which allows extreme pressure and flares large amounts of hair. Switch to 6/0 for tails, wings, and hackle.

As a matter of aesthetics and tying ease, thread should match the dubbing color, or the overall fly color. When thread and dubbing colors match, dubbing a thin body is much easier because less dubbing is needed to cover a like-colored thread.

Dubbing

Dubbing is a primary ingredient in many flies. Some is natural (rabbit or muskrat), but most is synthetic nylon. Natural dubbing is available either on the skin or blended. Synthetic dubbing is usually blended. Blended means the material has been chopped in a blender and is loosely matted.

The word *dubbing* has a dual meaning. Dubbing (the act) is applying any material directly onto and covering a single strand of thread that then is wrapped around the hook to help form the fly. Natural fur and blended synthetics are most commonly used, but any soft, smooth material that can be dubbed is referred to as dubbing (the material).

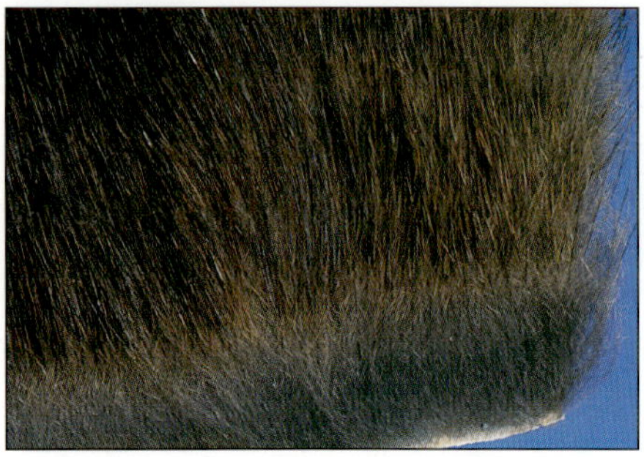

Muskrat is easy to dub. Pull guard hairs out for dry flies. Include them on nymphs.

Hare's mask is used for tails, abdomens, and thoraxes.

Blended squirrel body is similar to blended hare's ear.

Goat is often used in blends. It is a mainstay for nymph tyers.

Custom blend of several colors of goat and a dominant color of HareTron.

Important qualities to consider include sparkle, diameter, and texture. Texture is an important consideration, especially in regard to animation qualities and fly size. Smaller flies (16-20) usually need to be more exact and dubbed more tightly with a finer-diameter dubbing. Larger nondescript impressionistic nymphs (4-12) can often be more animated and shaggy in appearance and dubbed less tightly with a larger-diameter dubbing.

Natural dubbing fur on the skin usually has an abundance of guard hairs that may extend beyond the softer underfur. For example, muskrat and hare's ear have dense underfur with longer, darker, guard hair. When dubbed, the guard hairs protrude from the fly at odd angles, creating a natural-yet-abstract buggy, animated appearance. Synthetics lack guard hairs but offer various textures, almost unlimited colors, and superior sparkle. Some commercially-blended dubbings are a combination of both natural and synthetic materials. Most patterns are tied with synthetic dubbing that is available off the shelf.

Natural Dubbing

Hare's Ear is a popular and effective natural dubbing material used to construct the famous Gold-Ribbed Hare's Ear. Fish seem to be attracted to its natural characteristics. Hare's Ear dubbing is obtained from the face of a European hare, called a hare's mask/ears. Packaged preblended hare's ear dubbing is available, but I like to blend it myself. Trim the ears closely, then the face, except for the longer, lighter colored fur along the edges. Save a patch in the center for tails. Squirrel body is of similar texture and also makes wonderful nymphs.

Angora Goat is a natural, somewhat coarse, slightly crinkled fiber that can be used as is or blended with other dubbing. Because of its coarser nature, it is a bit more difficult to dub. Dub it loosely, wrap in place and adjust with fingers, bodkin, and scissors.

Rabbit Fur Strips on the hide are popular for tying leeches. Get the cross-cut rabbit. When it is wrapped around a hook, it angles back at 45 degrees. If you are using rabbit strips for a tail or Matuka-style fly, get the regular cut. It stands straight out from the skin and rides upright in the water.

Synthetic Dubbing

Antron has a high degree of reflectivity and is the most popular and widely used dubbing material. It is available in several forms, including dubbing, single-ply yarn, four-ply yarn, and "pure" silvery-white fiber strands. It is all of value.

Antron Dubbing is smooth, sparkly, easy to work with, and comes in an excellent selection of both fluorescent and natural colors. Antron makes fantastic dry flies and is excellent for nymphs and streamers when a smaller, tighter abdomen/body is needed. It is the all-purpose favorite.

Rabbit fur strips are natural rabbit on the hide.

Antron dubbing is perhaps the most useful. It has great sparkle and ties easily.

Antron Fibers are long, silvery white, and sparkly. They are used to represent a sheath-like air bubble (wings or tail) or to create highlights. Antron fibers are most commonly used in a blend with other dubbing materials to create a beautiful sparkle effect. Examples are rabbit or squirrel and Antron fibers, or hare's ear and Antron fibers.

HareTron is a blend of natural rabbit and sparkly Antron fibers, which offers translucence and ease of use. It is available in good colors and blends easily. This is a favorite all-purpose dubbing for nymphs and some dry flies. The many short, fuzzy hairs in HareTron accentuate and enhance dry fly bodies, but any long guard hairs should be removed on smaller, more delicate surface imitations.

SLF means "synthetic living fiber." SLF is slightly smoother in texture than angora goat and has more sparkle. Blend it for highlight sparkle—the visual effect is very pleasing. A smooth version similar to Antron dubbing is called SLF Finesse, which makes beautiful small nymphs and dries.

Plumage

Pheasant mentioned in this book is ringneck. Nymph tyers commonly use ringneck tail and body feathers. *Tail feathers* are used for tails, legs, wingcases, abdomens, thoraxes, and legs. It is easy to work with. The most popular colors are natural brown and dyed olive or green. The Pheasant Tail, page 46, is the most popular nymph tied with pheasant tail fibers. See the March Brown, page 68.

Peacock herl is the magic material that fish (and anglers) never reject. *Eyed tails* are the most useful. *Peacock swords* are the shorter, eyeless tails that have dense metallic herl one-half to two inches wide. Unless otherwise stated, whenever pattern dressings in this book call for peacock, you should use the herl on the eyed tail. The best herl is found on complete eyed tails immediately below the eye. Strung herl is gathered from the complete tail and includes the best and worst herls.

Partridge is one of the most natural looking and animated materials and is usually used for tying nymph legs and tails and some dry fly downwings. The fine diameter fibers are intricately mottled. It is an excellent substitute for other mottled materials of similar application (grouse). Some feathers are predominately brown, some gray, some mixed. For ease of feather color and size selection and to obtain a good selection of feathers, partridge is best purchased on the skin. Some popular nymphs tied with partridge include the Soft Hackle (page 43) and Copper John (page 46.)

Marabou has wonderful lifelike, animated qualities and is one of the most important materials for nymph and streamer tyers. Seven of our demonstration patterns call for marabou. Marabou is mostly used for tails, bodies, gills, wings, and wingcases.

HareTron is an excellent mixture of natural rabbit and synthetic Antron fibers.

SLF makes excellent "buggy," sparkly nymphs, especially leeches and stoneflies.

Natural ringneck tail and dyed body feather. Many flies call for these materials.

Peacock-eyed tail. The best herl is immediately below the eye.

Marabou is useful for all styles of flies. Mini marabou (left) and Woolly Bugger-sized marabou. Larger plumes are available.

Dark (brown) and light partridge. Many nymphs and some dry flies call for this.

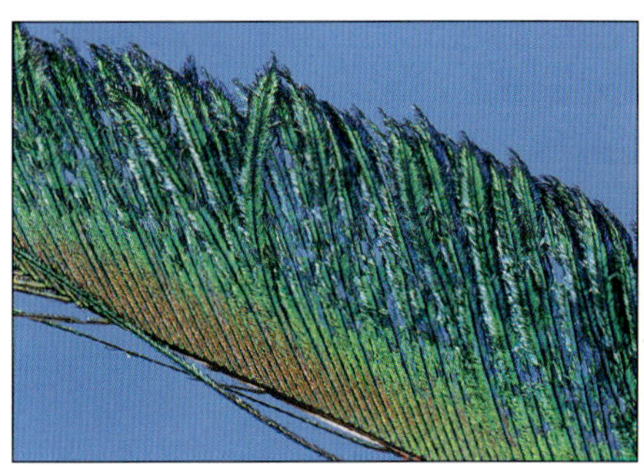

Peacock sword. The Zug Bug is one of only a few trout flies that call for peacock sword.

Natural turkey wing quills are scarce. They are used for Muddlers and hopper wings.

White tip turkey tail secondaries. Stiffen with lacquer or Flexament.

White turkey primary wing quill dyed brown. The short side has biot quills.

Natural teal flank (top), natural mallard flank, and mallard flank dyed woodduck color.

Turkey feathers are used in many flies (nymphs, dries, streamers) for wingcases, abdomens, tails, legs, wings, and antennae. There are four types of turkey feathers:

Natural wing quills are prized for tying Muddler Minnow and hopper wings. They are from brown turkeys (scarce) and are mottled brown with scattered white flecks. Most nymph tyers prefer the darker, less expensive tail feathers.

Turkey tail feathers are white tipped and mottled a dark brown. There are two types: the "secondary" feathers are about six inches long and soft, with fine diameter fibers. "Primaries" are about 10 inches long and stiffer, with more coarse fibers.

Dyed turkey wing quills (from Thanksgiving turkeys) are commonly used for backs and wingcases and are often preferred when a nymph pattern calls for goose.

Turkey biot quills make perfect V tails, antennae, legs, and bodies (Copper John, Prince). The wide side is usually too coarse for backs and wingcases. Turkey biot is used when a pattern calls for "stripped goose" or "goose biot."

Note that there is a "left" and a "right" wing feather on *all* bird wing quills (duck, goose, turkey, etc.). The widest side denotes whether it is a "left" or "right" feather. Dry fly tyers trim a section from a "left" and a "right" feather to form an upright and divided quill wing.

Mallard is the most commonly used duck feather. Mallard flank feathers, which are located along the sides of the bird, are incorporated into wings, tails, legs, and antennae. The natural feathers are barred dark gray and white. The most common dyed color is woodduck (lemon yellow).

Teal flank feathers have a pronounced black and white barring. Seldom used for nymphs, they are excellent for tying dry fly parachute wings.

CDC (Cul de Canard) are found by the oil preen glands of ducks. Because of their location, they are well impregnated with a duck's water-repellent oils. There are only a handful of feathers per bird, and the best come from wild ducks. Only prime feathers are of value.

Prime CDC feathers have *dense* fibers. It almost looks like duck down. The barbules of most other bird feathers extend at a right angle from the fibers. This allows the fibers to "marry" or adhere to one another. CDC barbules are unique in that they *spiral around* the fiber stem. This prevents them from clinging together and creates space between the fibers. All fibers react in an independent manner which creates air pockets and a uniquely animated light pattern and surface imprint. CDC should not be treated with floatants.

CDC is useful for tying emergers, legs, wings, wingcases and for "hackle" on dry flies. See the Mirage™ (page 44). The key factor is obtaining prime CDC.

Poor large-sized CDC (left). Stem is thick, fibers sparse; this is junk. Puff on right is excellent quality, fine stemmed with dense fibers.

Medium-sized CDC. Left feather is far superior to right. Note the difference in fiber structure.

Excellent CDC. Both puffs and larger feathers are used for nymphs and dries.

Reflective Materials

Krystal Flash is a spiral-wound, fine-diameter mylar offering beautiful sparkle and reflection. It has more body than Flashabou. It is useful for backs, wings, and as highlights in wings and tails. See the Matuka, page 53, and Woolly Bugger, page 61.

Flashabou is a strong, fine, flat, pliable mylar packaged in bundled strips about six inches long. Its reflective properties are amazing. Holographic colors are dynamite! It appears on many nymphs, streamers, and steelhead flies, mostly as backs, wingcases, and wings. See the Copper John, page 46.

Mirror Flash, the narrowest mylar, reflects three times more light than any other. Use for wings and highlights. See Hot Butt® Steelhead, page 57.

Tinsel and Wire

Wire is used for bodies and ribbing. It creates segmentation, adds some flash and a little weight, and is a great reinforcer for fragile materials such as peacock. Wire comes in several sizes. Unless otherwise noted, all dressings that call for wire rib refer to *fine* diameter wire. Larger diameter wire makes excellent abdomens (shiny and segmented) and aids in sinking a fly. See the Mirage™ page 44, and Copper John, page 46.

Flat Tinsel (mylar) is used as ribbing and is gold on one side and silver on the other. It is also available in pearl and holographic colors in various sizes (widths). Oval tinsel has a thread core and is also called rope, round, or braid. All tinsels are used as a reinforcer, to add sparkle, sometimes segmentation, and to create abdomens, backs, ribbing, and wingcases. See Marabou Muddler, page 55, and Green Butt Skunk, page 58.

Miscellaneous Materials

Non-Toxic Tin or ***Lead Wire:*** This is used to add weight to flies and is explained in the Weighting Flies chapter, page 29.

V-Rib (vinyl rib) is a soft, pliable, stretchable, half-round, small-diameter solid plastic that is perfect for forming and overwrapping abdomens on small hooks, especially midge nymphs. It is a substitute for Swannundaze.

Micro Fibetts are synthetic fibers with tapered ends for tails. See Parachute *Baetis,* page 78.

Yarn, Floss, and Chenille

Yarn is seldom used because dubbed flies are preferred. *Glo-Bug Yarn* is a smooth, soft, dense yarn that is usually used for tying Hot Butt® tails. *Leech Yarn* and *mohair* are sometimes used for tying leech imitations.

Antron Yarn is smooth, highly reflective, and similar to Z-Lon and Zelon. It is incorporated into many smaller nymph tails, veils, antennae, and shucks and is used to create sparkle highlights. Antron Yarn is often substituted for Z-Lon.

Krystal Flash is indispensible for highlights on subsurface flies and even some dry flies.

Holographic Flashabou and Mirror Flash are on "must have" lists.

Ultra Wire is colored copper wire available in many colors and sizes. It is very useful.

Flat (left) and oval tinsel is used for ribbing on all styles of flies.

Chenille comes in many colors, sizes, and textures. New Age, Crystal, and Vernille shown.

Antron Yarn (spooled or carded) is used for wings, tails, shucks, veils, etc.

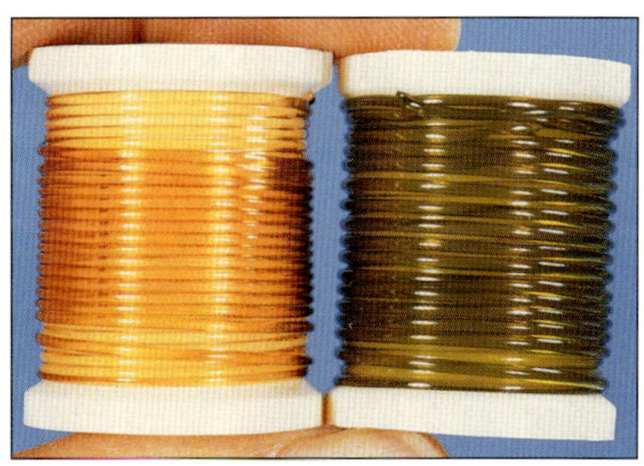

V-Rib (vinyl rib) is a substitute for and has replaced Swannundaze for smaller-sized flies.

Beads, coneheads, and eyes are best organized in plastic boxes with a curved bottom.

Grizzly color hen neck (top) and rooster neck.

Grizzly color saddle patch, showing hackle length. Saddles are a pleasure to tie with.

Deer hair, like elk, comes in assorted natural and dyed colors and textures.

Z-Lon and ***Crinkle Zelon*** are stiffer than Antron and vary in fiber structure. Popular for shucks, tails, wings, and highlights.

Four-strand Rayon Floss separates easily and is sometimes needed for ribbing or abdomens (bodies). See Soft Hackle, page 43.

Chenille is like a pipe cleaner without the rigid core. It comes in many colors and sizes and is used for tying abdomens, as on the Green Butt Skunk. *Sparkle Chenille* has a woven silver core that adds sparkle. *Ultra Chenille* is smooth, relatively stiff, and dense and is used to tie San Juan Worms. *Crystal Chenille* and *Ice Chenille* are pearlescent Flashabou-type mylar chenilles. They have many mylar pieces projecting from their core, which offers tremendous sparkle. They make excellent Woolly Buggers. See Misfit™, page 62.

Beads and Eyes

Beads are used on many styles of flies. They are available in several colors, sizes, shapes, and weights and are usually made of tungsten or brass. They are plated with copper, nickel, or gold. Coneheads (cone shaped beads) are also available. See the Weighting Flies chapter, page 30.

Hourglass-Barbell Eyes are shaped like mini barbells and are tied at the front of the fly to portray eyes, to add weight, and help create an up-and-down swimming or jig action.

Monofilament Eyes are burned monofilament ends. They make realistic nymph eyes. Buy them packaged. See Marabou Damsel, page 41.

Hackle

For our purposes, the term *hackle* denotes domestic chicken feathers, either rooster or hen. There are neck and saddle hackles. Many brands of genetically produced hackles are available, including Metz and Whiting. *Hackling* is the process of winding hackle around the hook. It is intended to represent tails, legs, gills, and wings, and creates animation, silhouette, and special effects. We discuss three styles of hackling, standard, palmered, and parachute. See page 48.

Hair

Hair is mostly used for dry fly and streamer wings and dry fly tails (Elk Hair Caddis, Clouser, Adams), overbodies (Humpy), and spun bodies and heads (Marabou Muddler). Hair is sometimes used on nymph-emerger patterns.

Briefly, there are several types of hair, including deer, elk, moose, calf, and caribou, all of which are selected and used for specific purposes. Considerations include stiffness, length, density, color, and buoyancy. *Hollow* is the term used to describe hair that has air pockets in it. Softer hair usually has more air pockets. The softest hair (caribou) flares easily and packs densely. *Moose* is the stiffest. *Caribou* is used for

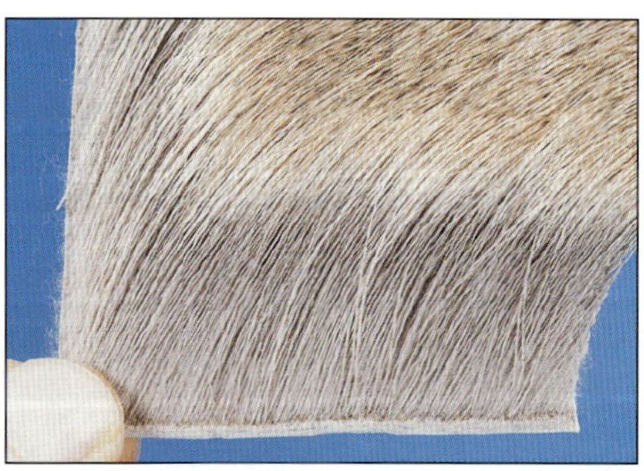

Elk hair is preferred for most hairwings on dry flies. This is light-colored elk.

Moose body is straight and stiff. It is mainly used for tails, legs, and overbodies.

Caribou is used for spun and trimmed bodies. It is too soft for wings, tails, etc.

trimmed bodies and heads, and moose is used for tails. *Elk* is a favorite for tails and wings.

Calf tail hair is used on many flies, usually for dry fly and streamer wings. See Green Butt Skunk, page 58. Select tails that have fine straight hair. Calf body is perfect for dry fly wings if you can find it long enough. Body hair is much finer and more straight than tails.

Bucktail from white-tail deer is popular for streamer wings, which are also called bucktail streamers. See Marabou Muddler, page 55.

Squirrel tails (natural gray, red fox, and dyed colors) are used on many classic streamer patterns. It is a fine, "hard" hair one to two inches long that is beautifully barred. See the Clouser Minnow, page 54.

Most hair flies are considered advanced and beyond the scope of this book. Tyers interested in a thorough discussion of hair and dry flies that incorporate hair, such as the Humpy, Irresistible, Wulff, and Stimulator, should consider my book, *Tying Dry Flies,* and the Jack Dennis book series, including *Tying Flies with Jack Dennis and Friends* and his many videos, including *Tying Western Trout Flies, Tying Western Dry Flies, Tying Flies for Spring Creeks and Tailwaters,* and *Tying Cripples and Spinners* with Kelly Gallup.

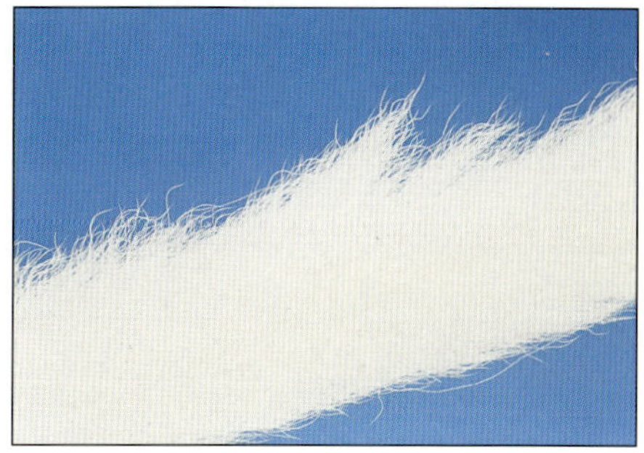

Calf tail is popular for tails and wings. Use calf body when possible for dry flies.

Squirrel tails (gray and red fox) are used mainly for streamer wings.

Material List

The following materials cover almost every fly pictured in this book, and probably a hundred others. Once you have a collection of material like this, you can do some serious late-night tying!

Hooks
900BL: sizes 12, 14, 16
3761BL: sizes 12, 14, 16
200RBL: size 10
5263BL: sizes 6, 8, 10
2457: sizes 12, 14, 16
2487BL: size 18
7999: sizes 2, 4, 6

Beads
Metal: gold, copper, size 3/32", 1/8", 5/32"
Tungsten: gold, copper, nickel/silver, sizes 5/64", 3/32", 7/64", 1/8"
Cone: gold, size 1/4"
Glass: amber, small

Eyes
Burned monofilament: extra-small, small
Barbell: painted yellow/black size 6/32", nickel plated size x-small, small, medium

Wire
Tin: sizes .015, .025, .035
Ultra Wire: copper (sizes small, brassie, medium); green, red (size brassie)
Fine wire: gold

Thread
6/0: red, black, brown, maroon, olive, fluorescent fire orange, fluorescent yellow, tan, gray, pale yellow

Dubbing
Antron: black, golden olive, medium olive, golden tan, olive damsel, mint, light gray, charcoal gray, Cahill cream, March Brown
HareTron: black, yellow, burnt orange
Ice Dub or *STS Trilobal:* black, purple
SLF: red
Hare's Ear/mask: natural
Opossum: natural

Yarn, Chenille, Floss
Antron yarn: cream, white, golden yellow, olive, green
Glo-Bug yarn: Alaskan roe, cerise
Chenille: fluorescent orange, green, pink, fire orange, size small
Chenille: black, olive, purple, sizes small and medium
Ultra Chenille: red
Crystal Chenille: pearl, pearlescent black, and purple
Chenille: Glo-Brite, black
Floss: yellow, black, olive
Fuzzy Wool: fluorescent fire orange

Hair
Deer: medium soft and medium hard natural brown
Elk: natural light , natural gray
Moose body
Caribou
Squirrel: red fox, red fox dyed yellow
Calf tail: white, dyed brown
Calf body: white

Plumage
Turkey biot: brown, white
Peacock: tail, sword
Partridge: brown
Marabou: maroon, olives, browns, tans, black, white, cerise, hot orange, purple
CDC: puffs and large select, natural gray
Mallard: natural and dyed woodduck
Pheasant tail: natural
Turkey: white-tip secondary

Reflectives
Krystal Flash: pearl, peacock, black, olive, gold, red, green, cerise
Flashabou: pearl, holographic silver, gold, red, black
Holographic Fly Flash: silver, gold
Mirror Flash: pearl, peacock black, blue, purple
Tinsel: flat mylar size fine, small, medium; oval silver, size fine
Holographic flat tinsel: gold, size med.

Hackle
Saddle: furnace, grizzly, blue dun, light ginger, for hook sizes 8-14
Hen saddle: black
Hen neck: blue dun, grizzly
Strung Saddle: extra-select, black, purple, cerise, red, fluorescent orange

Miscellaneous
Micro Fibetts: blue dun
Head Cement
Crystal Epoxy
Zap-A-Gap
Premium Wax

Chapter 4

Basic Tying Techniques

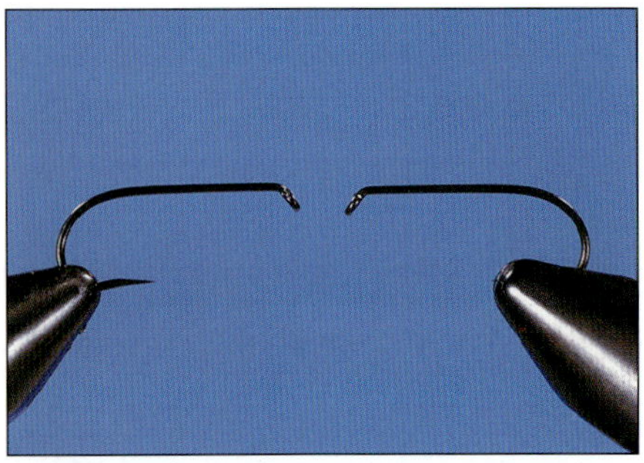

The hook on the left is properly inserted. The one on the right is not.

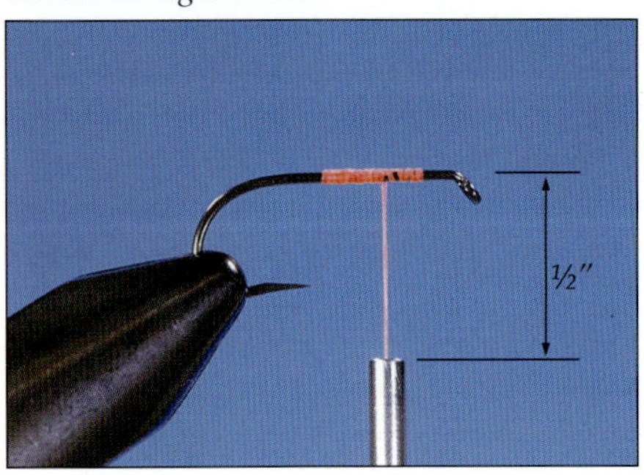

Once the thread is secured onto the hook (see next page) and you begin to attach or secure materials onto the hook with thread, there are a couple of efficient tricks to remember. When securing materials and wrapping thread, keep the distance from the hook shank to the bobbin tip to about ⅜ to ¾ of an inch. By doing this, a small-diameter circle is in play, allowing for quick and exact placement of thread.

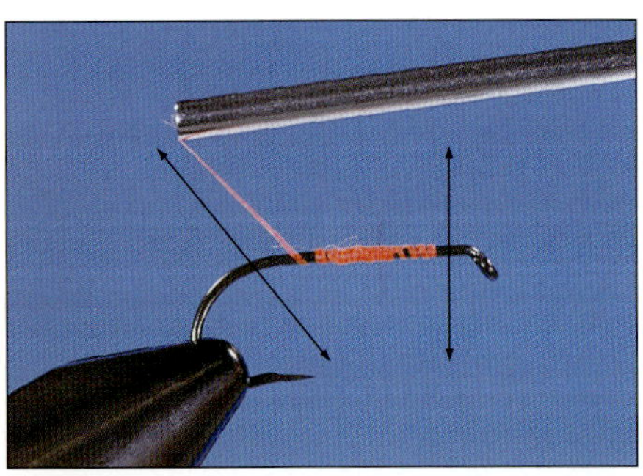

When working near the hook point, wrap thread at an angle to avoid hitting it and possibly breaking the thread. This takes a little practice, but, by turning a small circle, it soon becomes habit.

This chapter explains the basic tying techniques needed to tie flies. They should be fully understood and practiced before beginning actual fly construction. All directions are given for tying right handed. Left-handed tyers should turn the vise around and reverse all instructional directions.

Inserting The Hook

Inspect the hook carefully. Check the point, eye, shank, and hook temper. If there is a barb, smash it. Hooks should be mounted in the vise parallel to the table. Insert only the lower bend of the hook in the forward topmost part of the jaws. This allows you to maneuver and tie materials onto the hook more easily. The point and barb should be exposed because vise jaws can weaken a hook point. This is especially true of small, light wire hooks. Be careful not to catch your thread on the hook point. To avoid the point, tilt or angle the thread-wrapping circle.

Bobbin Operation

Before beginning, we need to thread the bobbin, adjust its tension, and learn how to operate it.

To adjust the tension, spread out or squeeze the ends of the arms together. Avoid putting pressure at the point where the arms are secured to the tube. Tension is properly adjusted when you can quickly pull out several inches of size 6/0 or 8/0 thread without breaking it. Tension is too loose if you need to continually roll up the thread spool or if thread unspools as the bobbin hangs from the hook.

Next, insert the wire loop threader through the tube toward the spool of thread. Pull six inches of thread from the spool and insert end of thread through wire threader. Pull threader and thread through shaft.

Rest the bobbin in the palm of your hand and grip the plastic pad with your thumb and first finger. Do not hold onto the round shaft because it rolls between your fingers or you have less control. Do not hold onto the thread. If you stop the spool from turning with your palm as you wrap, the thread won't feed out and the length of thread shortens. By easing off the tension and/or winding with more force, you cause the thread to feed out as you wrap it around the hook.

The top of the bobbin shaft should usually be kept close to the hook shank, which creates a small (¾-inch or less in diameter) circle when wrapped around the hook. The small circle allows you to *quickly* place thread *exactly* where you want it. I have a half-dozen bobbins pre-threaded so it is not necessary to re-thread one when I change thread colors or break the thread. A short shaft works better for short shank hooks and smaller flies. A long shaft is better for tying on long shank hooks.

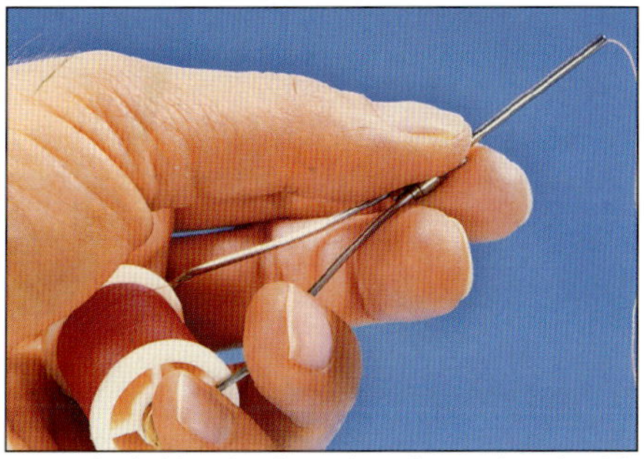

Rest the bobbin in the palm of your hand with your thumb and first finger on the plastic pad. Tighten your fourth and fifth fingers around the spool to apply tension.

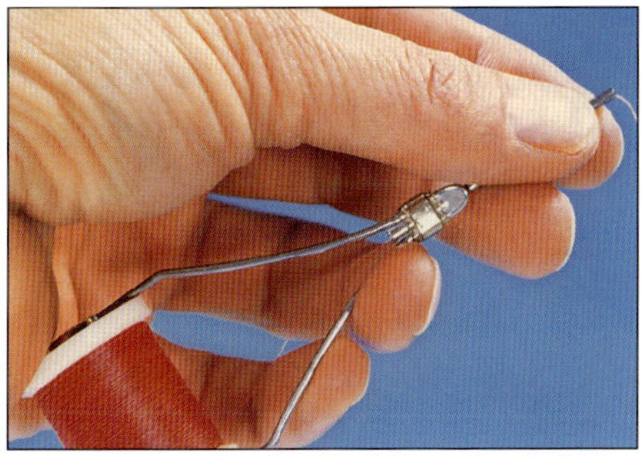

Incorrect method of holding the bobbin. Round shaft rolls between fingers, resulting in less control.

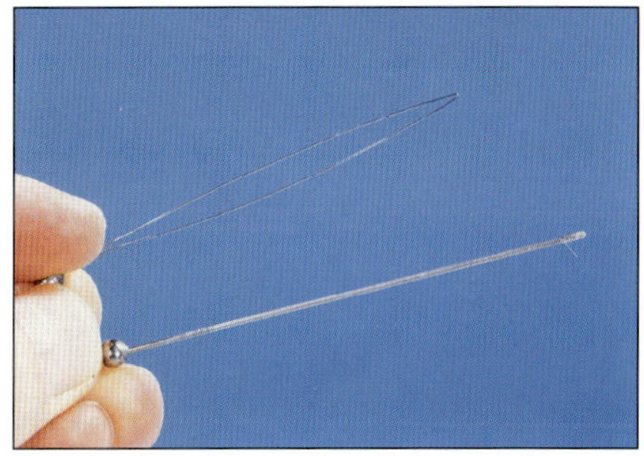

Bobbin threader (top) and cleaner (bottom) are connected by bead chain and are indispensible for cleaning the bobbin shaft and pulling thread through the shaft.

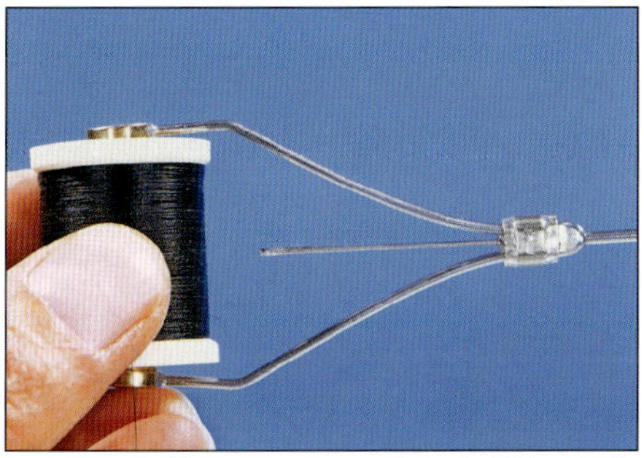

To clean built-up wax out of the bobbin shaft, insert solid shaft, pushing out wax.

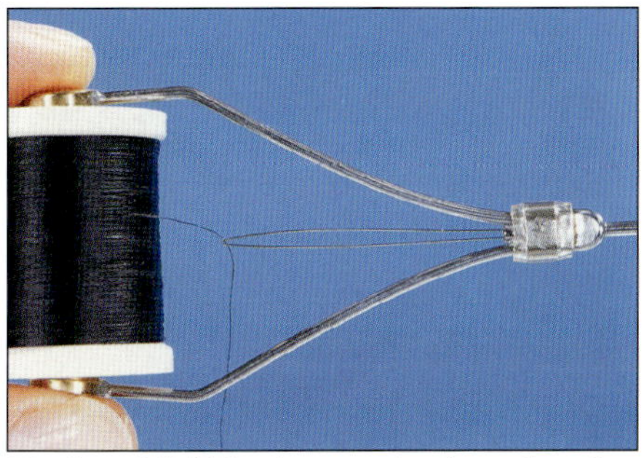

To thread bobbin, insert wire loop threader. Insert thread and pull through.

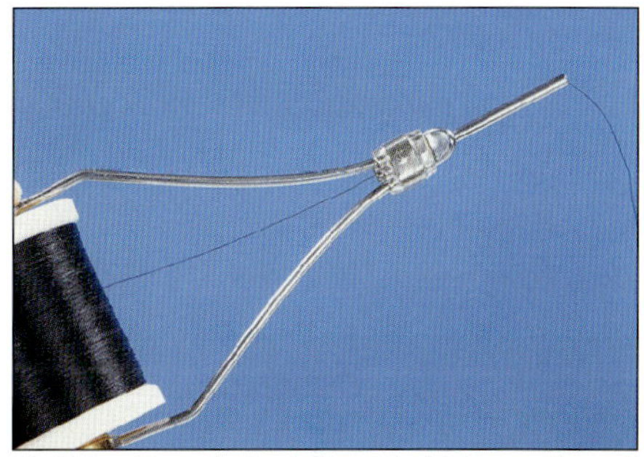

Bobbin threaded. Adjust bobbin so thread pulls off spool in line with shaft. This reduces wax build up and helps minimize thread breakage.

Securing The Thread

The function of thread is to secure material onto the hook. Thread should always be positioned at the *exact* location where material is to be secured. Whenever a material is secured onto the hook, it is referred to as *tie in, tie on, tie down* or *tie off.* Thread should always be wrapped in the same direction and as tightly as possible. The number of wraps should be limited to those that are absolutely necessary.

Thread is secured to the hook by doubling back, or wrapping, the thread over the top of itself, or over the short end. Hold the end of the thread in your left hand and place it against the hook. Keep tension on the thread and, with the bobbin in your right hand, wrap the thread forward in a clockwise direction around the hook, then back over itself.

Thread and materials should always be wrapped clockwise. Any time thread or materials are wrapped around the hook, *constant* tension must be applied. The idea is not to allow previously placed wraps of thread or material to unwind or loosen.

Thread breakage could be the result of exerting too much tension on the thread, chafing the thread on the hook point, or encountering a rough burr on wire, tinsel, etc. When the thread does break, hold on to the broken end with your fingers or hackle pliers and, if possible or necessary, half-hitch (see page 27) the broken end to keep it from unravelling. Rethread the bobbin, tie on the thread, and continue.

Hook shanks are round, smooth, and slick. If a foundation of thread is not wrapped over the hook before materials are secured, the finished fly may spin around the hook shank. When you begin, secure thread onto the hook shank and wrap it to the place where the first material will be tied in, or onto, the hook.

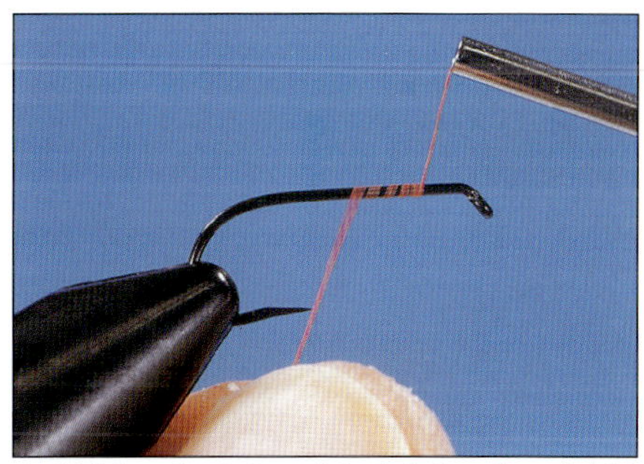

1. Hold end of thread in left hand and fold over top of hook, keeping it taut. Wrap three to five turns of thread forward (or in place) around the hook. I usually wrap thread in place—the forward direction is exaggerated here for visuals. Basically the thread needs to be wrapped back over itself.

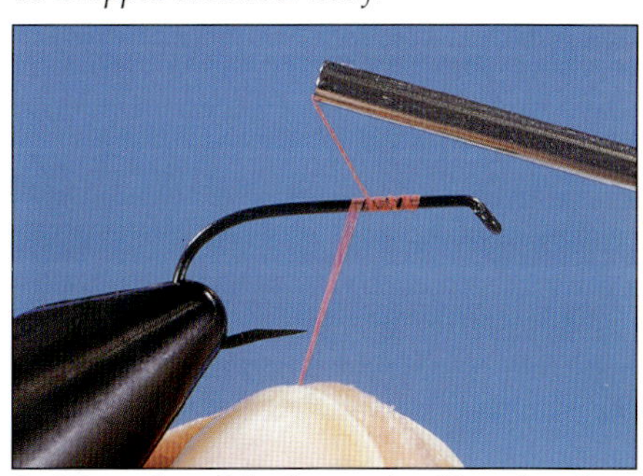

2. Wrap thread back over itself or the short end about six turns. Cut or pull off short end. Easy!

Securing Materials with the Up-Between-the-Fingers Pinch Technique

The up-between-the-fingers pinch technique is the best method to *quickly* accomplish a tie in or tie on at the *exact* position desired. This method prevents you from chasing material around the hook with the thread and eventually catching and tying it at an unintended location in a haphazard manner. Materials should be tied evenly *along* the hook shank, not *around* it.

Some parts of a fly, such as most tails and wings, are simply *secured in place* and are not wrapped around the hook. Materials that form an abdomen and hackle must be wrapped *around* the hook.

Tie a length of Antron yarn onto the hook to form a tail. Antron would only occasionally be used for a tail, but it serves to demonstrate the tie-in technique.

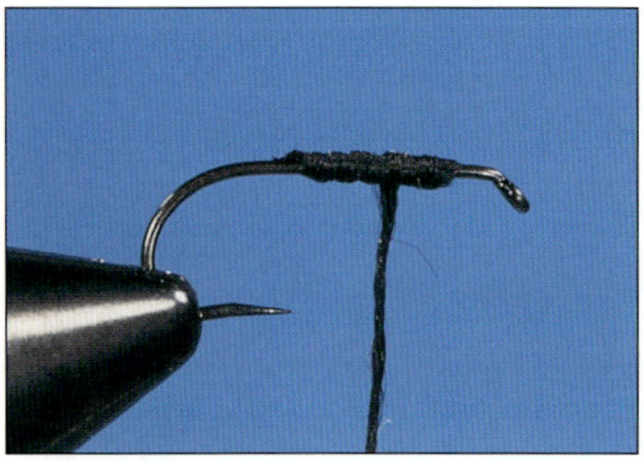

1. Secure the thread onto the hook. The thread is positioned at this position to tie in a tail because it will ensure a level or properly tapered underbody. Floss is used for visibility.

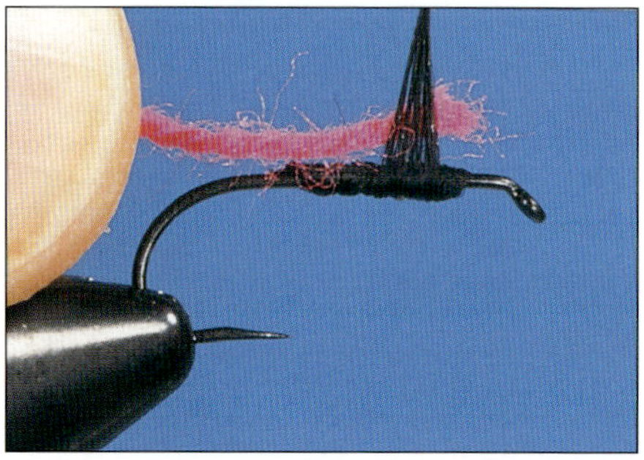

2. Cut a two-inch length of yarn and position it on top of the hook. Attempt to tie the yarn onto the top of the hook.

3. The thread pushes, or chases, the yarn around the hook.

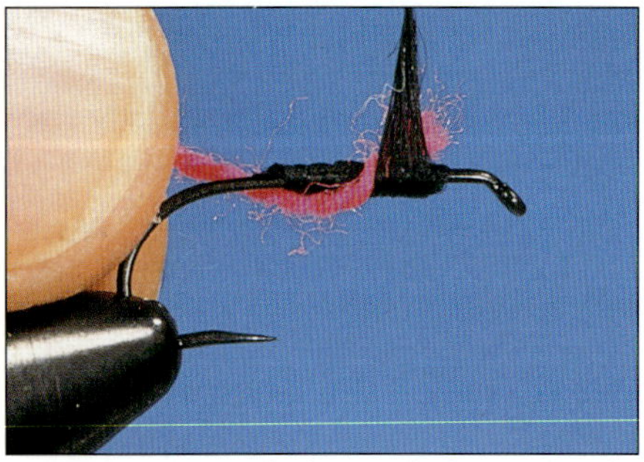

4. It may eventually be caught but not at the position you want.

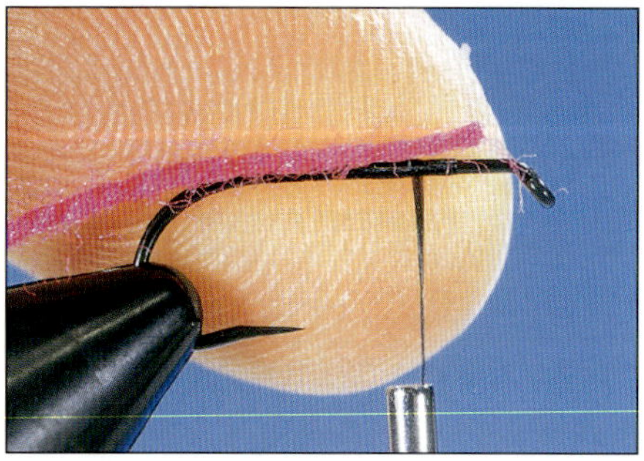

5. Position material to be secured on top side of hook. Place index finger along the side as pictured and then place thumb on the other side of the hook. See next photo.

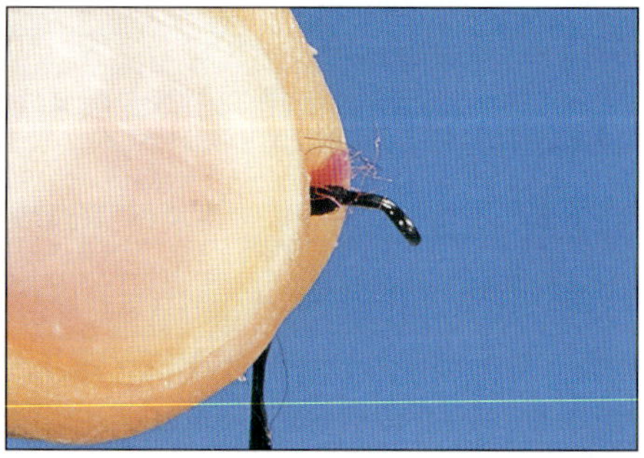

6. Thumb and first finger have pinched the yarn and thread and the hook. I usually pinch the material and set it in position, then secure.

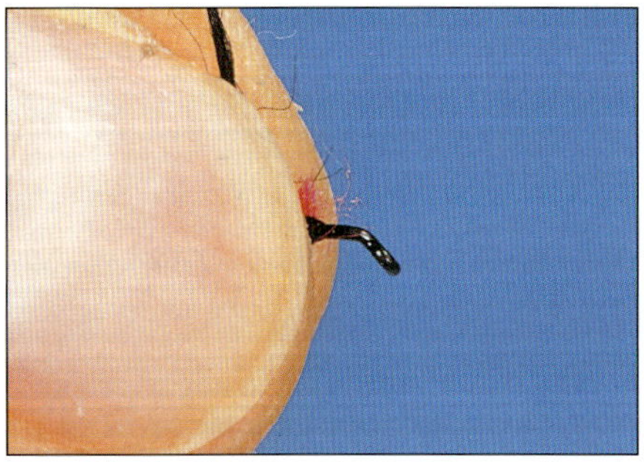

7. Bring the thread up between your fingers. Do not open your fingers to accomplish this. If necessary, bring the thread in front of your fingers and slide it between them.

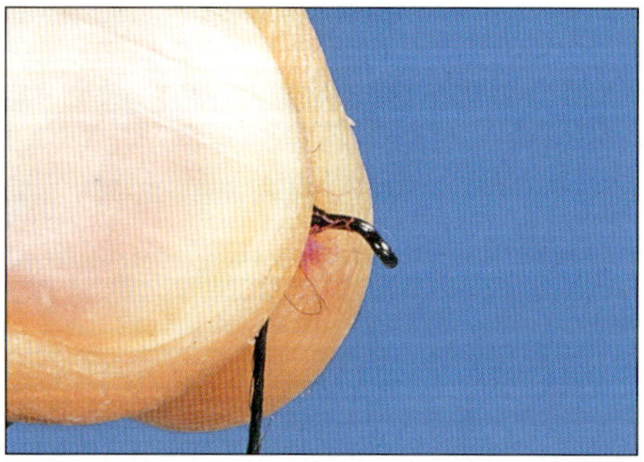

8. Once in position, bring the thread down over the opposite side of the hook, placing a turn of thread around the yarn. If the yarn is not positioned where you want it, reposition it and ***hold*** *it where you do want it.*

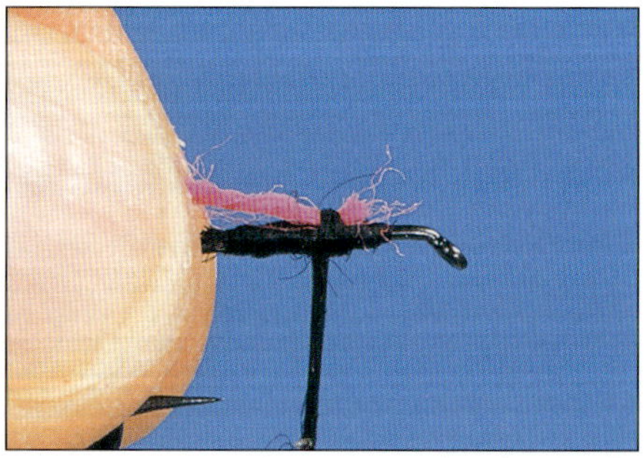

9. Yarn is properly positioned and initially secured on top of hook with one turn of thread. From this position, we would secure the yarn along the top of the hook, but, first, let's look at another tie-in technique.

Securing Materials with the Loose-Wrap-and-Cinch Technique

Often it is more expedient to take a loose (tensionless) wrap over the end of a material you want to tie in. Your first finger or thumb keeps the material in position while a loose, or tensionless, wrap of thread is taken over the material. When the loose wrap of thread is in place, it is cinched tight, locking the material in place. Place two to three more tight wraps to ensure the material is well secured. It's easy.

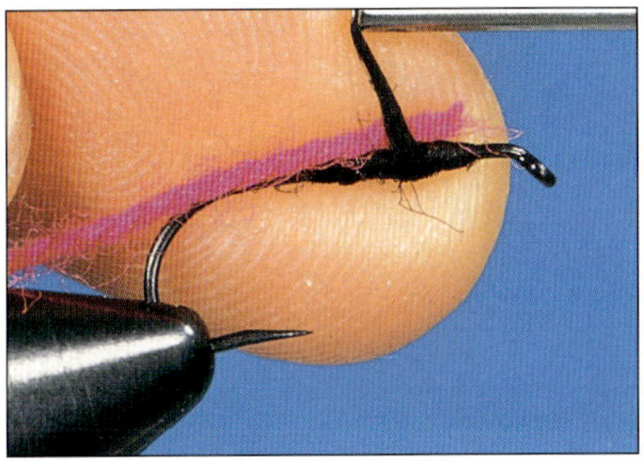

1. Position your first finger and the very end of the material to be tied in as shown.

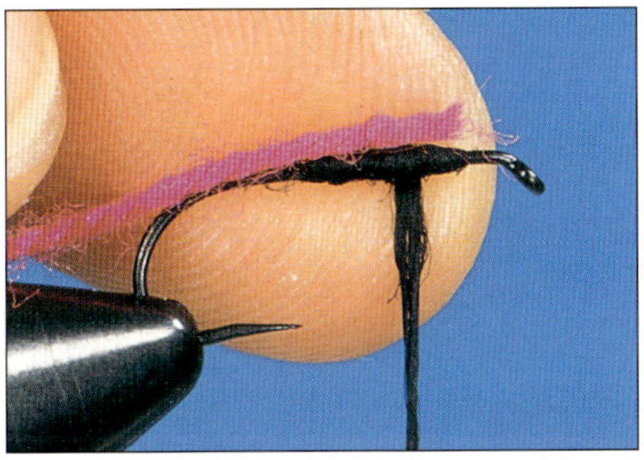

2. Your first finger keeps the material and previous turns of thread in place while a tensionless turn of thread is brought to the top of the hook.

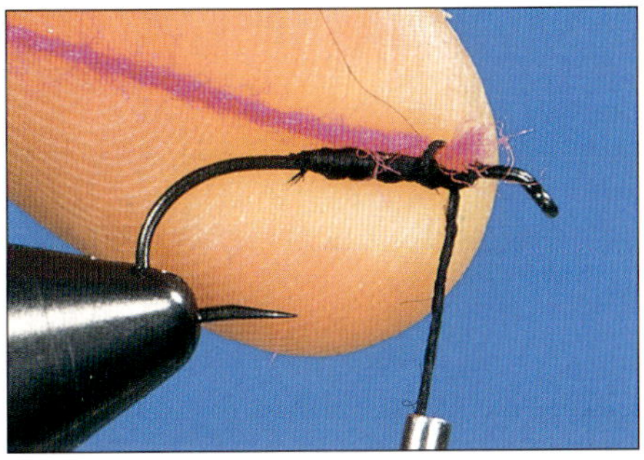

3. As you bring the thread down the opposite side of the hook, slide the thread between your finger and the yarn. Your first finger keeps the yarn in place. Pull the thread tight at the bottom of the circle around the hook.

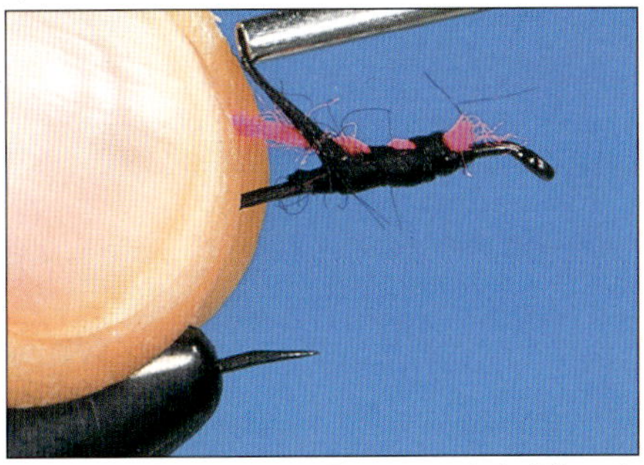

4. Pinch the yarn as shown, and, holding it at an angle above the hook shank, wrap the thread toward the back of the hook. Yarn (or any other material) usually should be secured straight along the top of the shank.

5. By doing this, a level or slightly tapered underbody is formed, ensuring a perfectly tapered finished fly. Secure the material back to a point just short of where the hook shank begins to curve downward.

Wrapping Materials with the Over-the-Top-and-Hold-It Technique

Tie in a length of floss, which becomes the abdomen (body). This is wrapped around the hook with the over-the-top-and-hold-it technique. When materials are to be wrapped around the hook, the thread must be repositioned to the tie off area after the material is tied in. The tie-off area is usually at the front of the part you are constructing; in this case, the abdomen.

Materials are usually wrapped back to front. The left hand does *all the wrapping*. The right hand only holds the material at the bottom position until the left hand comes back over the top of the hook, picks up the material, and begins another wrap around the hook. The right hand also acts as a bobbin rest, keeping the bobbin out of the way when material is being wrapped. Bobbin "rests" are available and may be of value to some tyers.

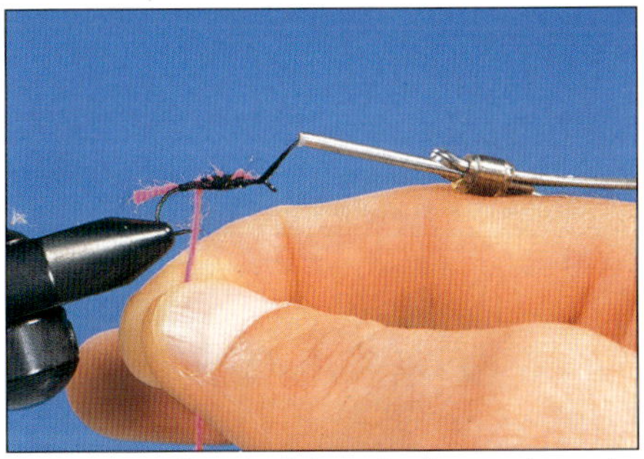

*1. Secure a four-inch length of yarn along the top of the hook back to the tail area. Rest the bobbin over the top of your right hand. Using your right hand, grasp the yarn with your thumb and first finger and hold it under the hook. This is the **hold-it** position.*

*2. Your right hand always stays in this position—under the hook. Your left hand does all the wrapping. Grasp the yarn with your left hand and make a 360-degree turn around the hook. This is the **over-the-top** movement.*

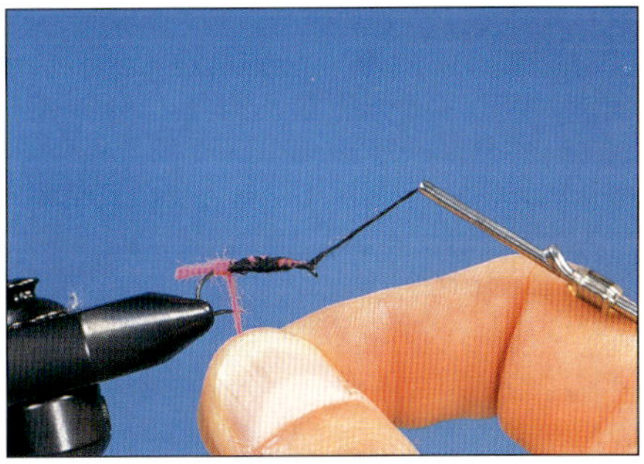

3. After a 360-degree turn has been accomplished, your right hand holds the yarn as shown until you bring your left hand counter-clockwise back over the hook, ready to make another turn. This is the left hand over-the-top, right hand hold-it technique. Materials are almost always best wrapped with this method.

Tying Materials Off

Once the material has been wrapped in place, it must be tied off, or secured onto the hook. Take two or three turns of thread over the material where you want it to end on the fly, tying it off. All material that is wrapped around the hook is tied off in this manner.

1. Hold the yarn (or any other material) in your right hand at the underside of the hook.

2. With your left hand, take a turn or two of thread around the hook at the tie-off position. The left hand brings the thread (bobbin) over the top and releases it, curving back over the hook to pick it up again.

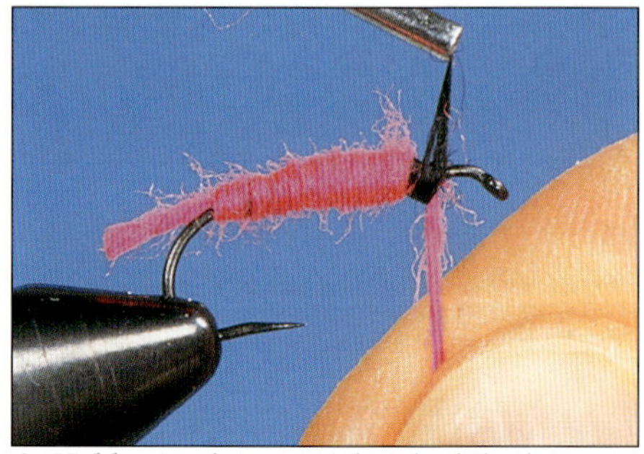

3. Hold onto the material end while these one or two turns of thread are being placed.

4. Release your right hand from the material. Pick up the bobbin with your right hand and make additional turns as needed to secure material (tie off). Remember to make tight wraps with the thread.

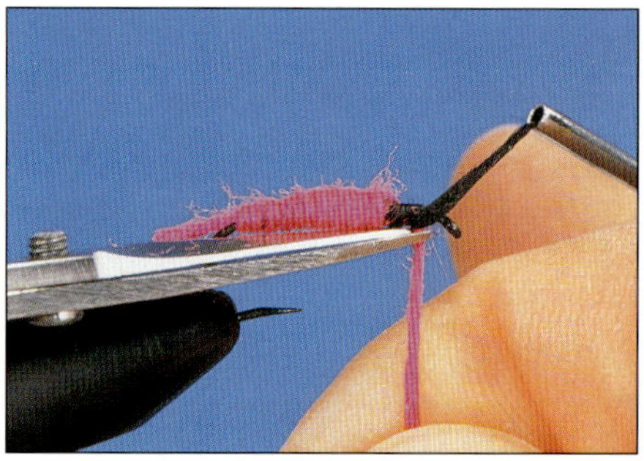

*5. Once secure, trim off the excess. To make a clean cut, hold the bobbin (thread) out of the trim area. Gather up **all** material to be cut and hold the ends taut. Place the tip of the scissors against the shank and make one clean cut.*

6. If done correctly, no materials, wild fibers, or stubby ends should remain.

Thread Heads

The thread head is where the fly is finished off with a knot just behind the hook eye, after which a drop of lacquer is applied.

Heads usually should be *small* in diameter, *smooth*, and *slightly tapered*. This is the hallmark of a finely crafted fly. A crowded head can preclude threading a leader through the hook eye. *Reserve a small space immediately behind the hook eye* solely for the thread head. This ensures enough room. Finish an average-sized nymph or dry fly head with about 10 wraps of 6/0 thread.

The thread head is the last step in completing a fly. Practice constructing it first so you can see how simple it is to tie a perfect head every time.

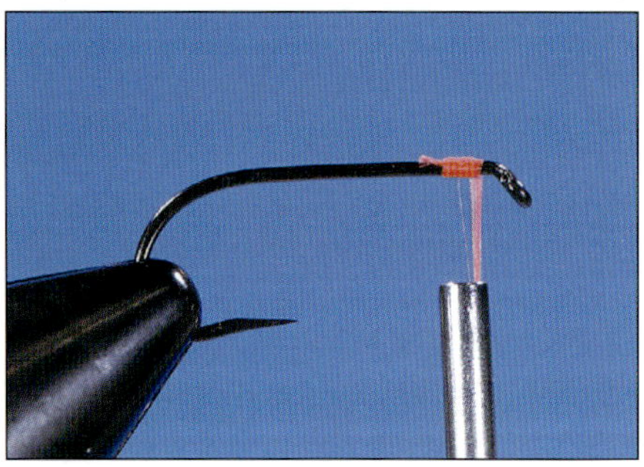

1. Pretend the fly is complete and the thread is at this position, ready to finish the head. Note the reserved space behind the hook eye.

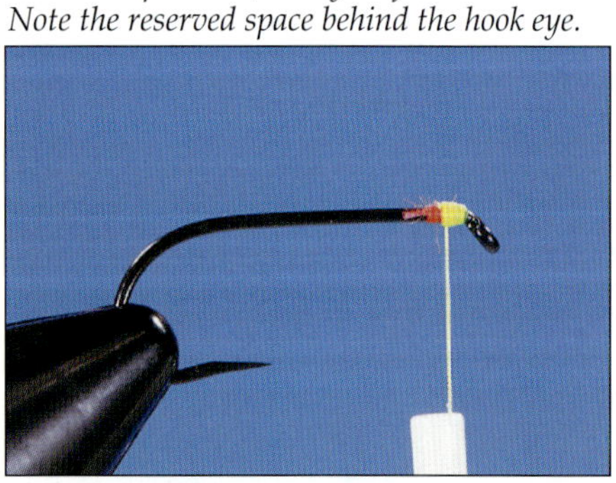

2. I changed thread colors to highlight the size and shape of the head. Thread is positioned at back of head. Tie two half hitches here, not at front of head.

*3. Nymph and dry fly thread heads should look like this. **Leave space**. Practice tying a neat head.*

4. If your head looks like this, try it again. Reserve space, keep all errant fibers out of the way, and use less thread.

Half Hitch

When properly executed, the half hitch is a quick, simple, neat, and effective knot for tying off the thread at the head of the fly. The half hitch is simply a loop of thread around the hook. I like to place *two* half hitches at the *rear* of the thread head area. (One half hitch does not hold, and three cannot cinch tightly.) Proper placement ensures a neat head. Before cutting the thread, cinch the two half hitches by pulling the thread counterclockwise around the hook (over the top from the back side). You will feel the two half hitches cinch together. Back off slightly on the tension and cut the thread close to the head so no loose end protrudes from the thread head. Lacquer the entire head.

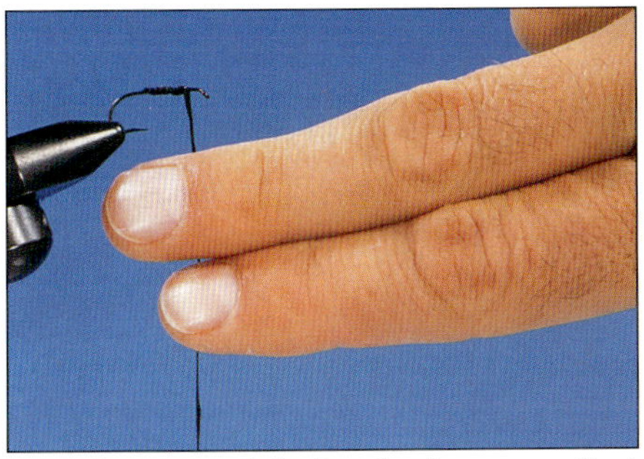

1. Using your right hand, place your first two fingers on the thread. Thread should be across the tip of your first finger. Your palm faces away. Your second finger keeps downward tension on the thread, and your left hand supplies tension on the bobbin as needed.

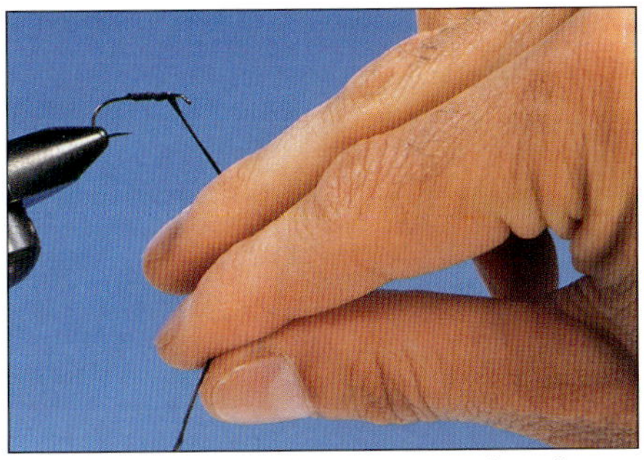

2. Bring your thumb up to your first finger, pinching the thread. Rotate your hand clockwise 180 degrees, forming a loop. The palm of your hand should be facing you. It is important that the thread is positioned on the tip of your first finger.

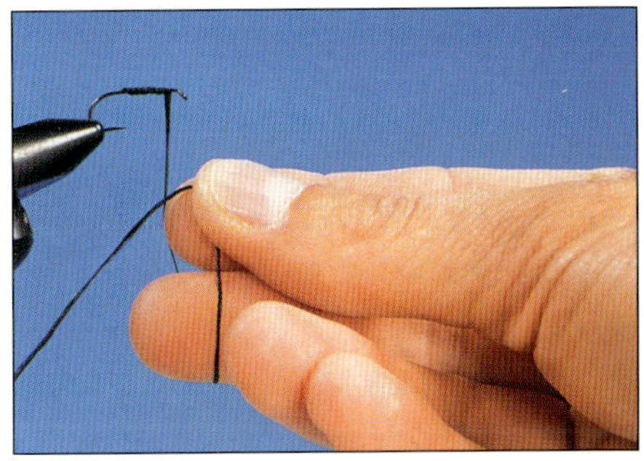

3. The loop is formed and is ready to be placed around the hook.

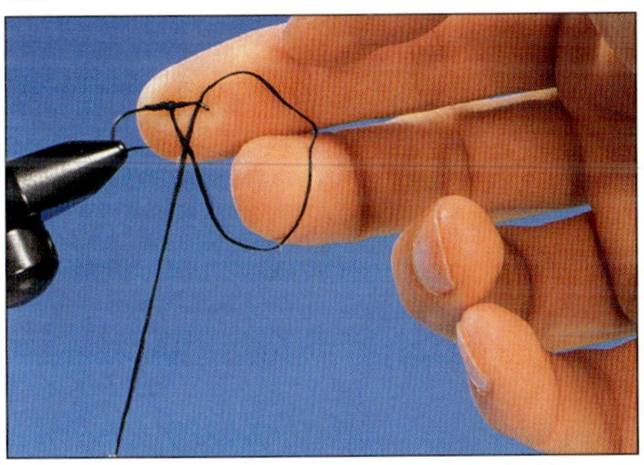

4. Place the loop around the hook, holding it in place on the far side of the hook with your first finger. Note the position of the thread across the first finger.

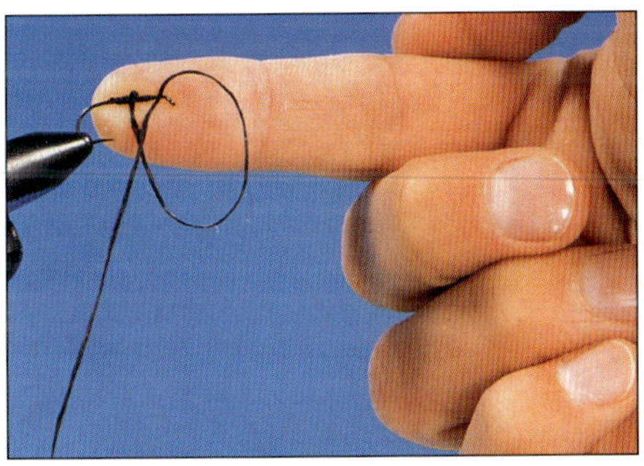

5. Remove your second finger from the loop. Keep your thumb in position as you draw the loop closed, which keeps the thread in front and away from the front of the fly. (See photo 4.) The thumb was removed for visibility in this photo.

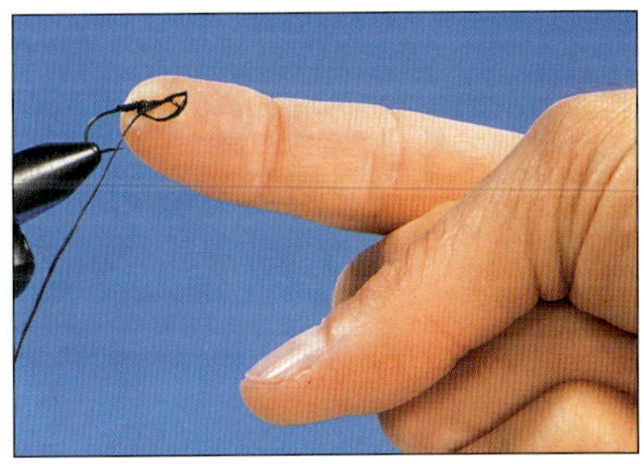

6. Remove thumb when the loop is nearly closed. The first finger stays on the loop until it is completely closed, ensuring proper placement. Place two half hitches, cinch them tightly counterclockwise, back off slightly on the tension, trim the thread, and lacquer.

Applying Lacquer

When applying lacquer, pick up a small drop on the *tip* of the bodkin and spread it over and around the head. The lacquer should be of the consistency that it adheres to the needle point and follows it around the thread head, slowly soaking in. If lacquer does not soak into the thread head slowly, it is too thick and should be thinned. If it is too thin, it saturates and does not build or cement the head well enough, or it may "run" back into the fly, glueing individual fibers, wings, etc. together. Lacquer can soak into the base of the hackles or legs but should not glue the individual fibers, wings, or guard hairs together. Keep lacquer out of hook eyes. It is frustrating to attempt to thread a leader through a hook eye that has been sealed shut, especially at dusk! Once lacquered, flies can be set down to dry. Epoxied flies should be turned so epoxy does not sag. Heat from your tying lamp helps speed the drying process.

Carefully "lead" a drop of lacquer around the head. If it seeps into the hook eye, clear it out.

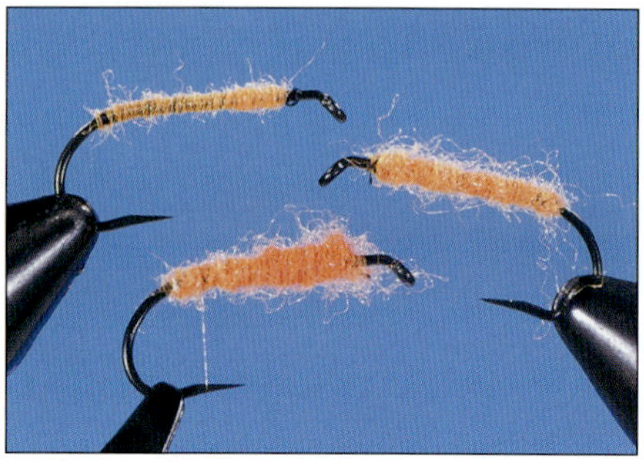

Light, medium, and heavy applications of dubbing. Black thread is used for visibility but shows through on the light application. Match thread to the dubbing color.

Dubbing

There are several dubbing (the act) methods. Direct dubbing, whereby dubbing (the material) is applied directly to thread, is the most common, easiest, and efficient. Smooth dubbing material, such as Antron and HareTron, is easy to dub tightly onto thread. The coarser the dubbing material, the more difficult it is to dub tightly. When dubbing a coarse material like goat, do a little at a time and do not try to dub it too tightly or it falls away. Do not concern yourself with the fact that it is not tight. When you wrap it onto the hook, overlapping wraps secure enough fibers, and any excess can be pulled out.

Practice dubbing the *least* amount of material—only changing the thread texture!

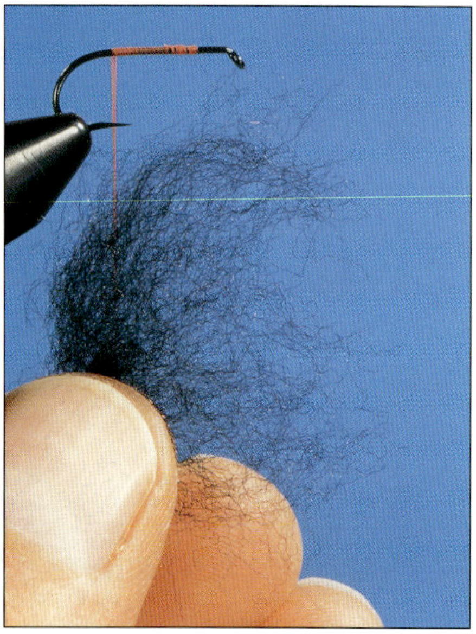

1. Begin with very little *material. Hold what you think you need between your thumb and first finger. It might look like this. Select 10 percent of what is pictured.*

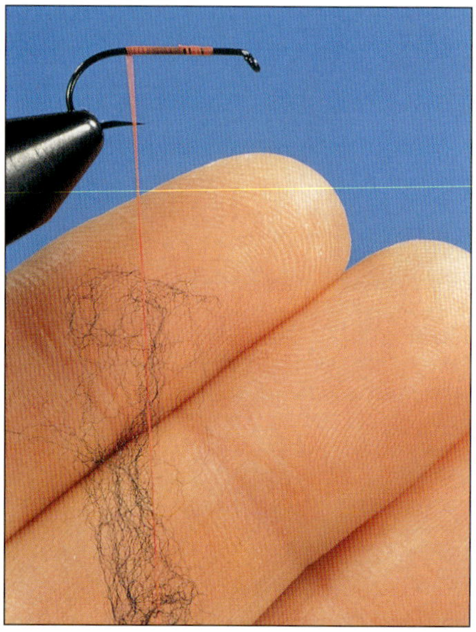

2. For flies size 16 and smaller, dubbing should only add color and sparkle, not bulk. Thread is bulky enough! Apply dubbing to no more than two or three inches of thread. Wrap into place and dub more as needed.

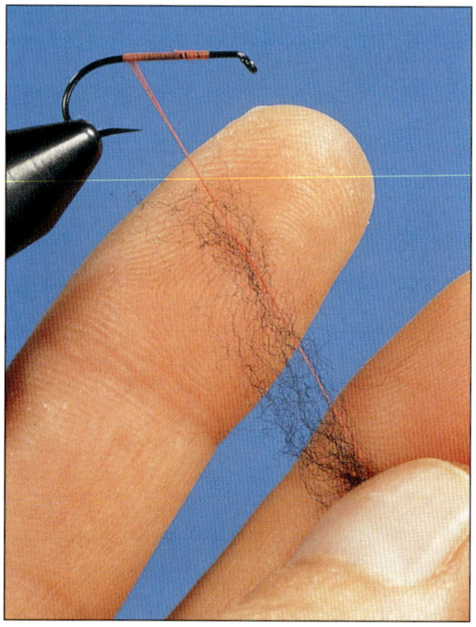

3. Spread dubbing fibers evenly and sparsely. Start dubbing three-fourths of an inch below hook point, dubbing the top-most fibers into place. Start at the top-most fiber and work down the thread.

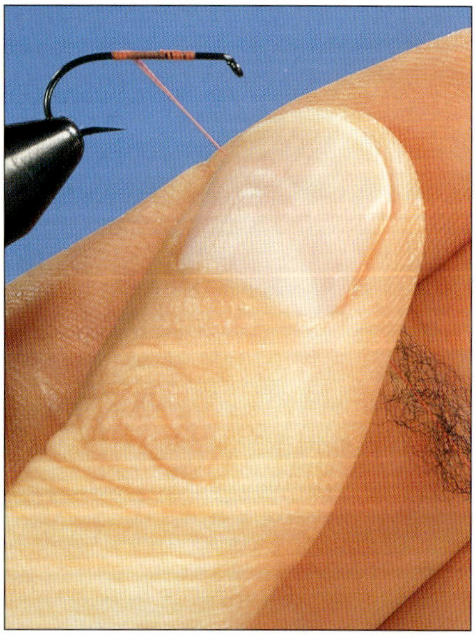

4. This allows excess dubbing to be squeezed out below. Keep bobbin thread taut. Roll fingers together tightly in one direction, under pressure. Thread rolls, too. Do not roll back and forth—this reverses what you just did.

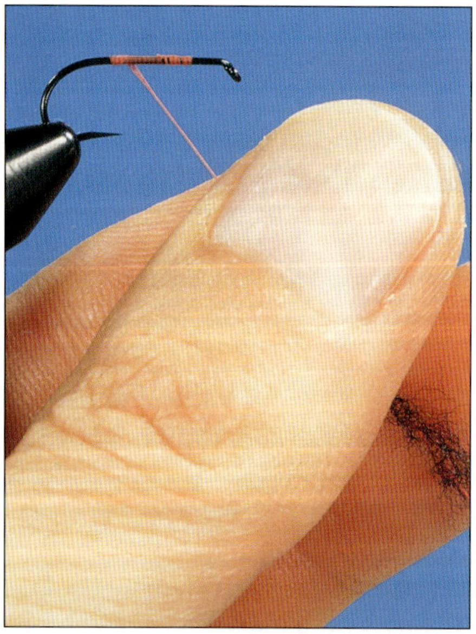

5. One roll (two or three at most) adheres material to thread. The actual "roll" begins at the first or second joint of fingers, ending with thumb at tips of fingers. Pressure is the key.

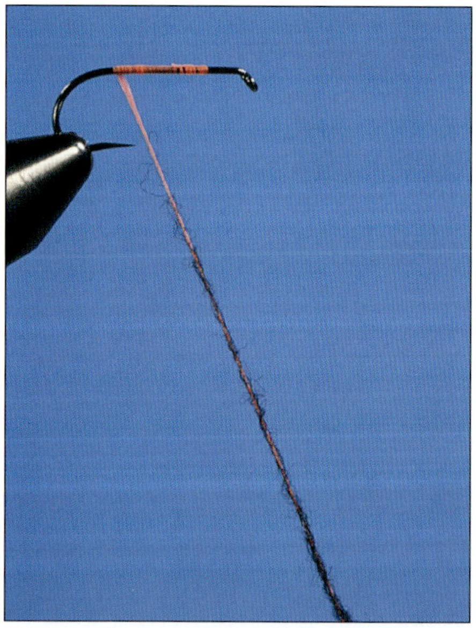

6. Properly installed dubbing. Notice that there are no undubbed or wild fibers extending from the top-most area of dubbing. Dub two to three inches at a time. Add as needed.

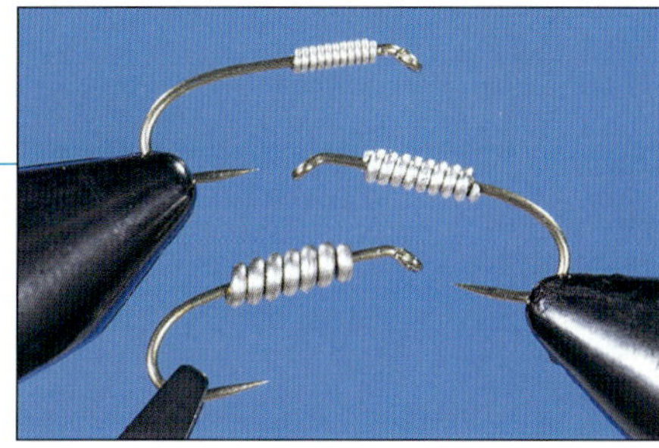

Add any amount of wire weight you like to hooks. Shown here are three possibilities for a 1X long Tiemco 3761BL.

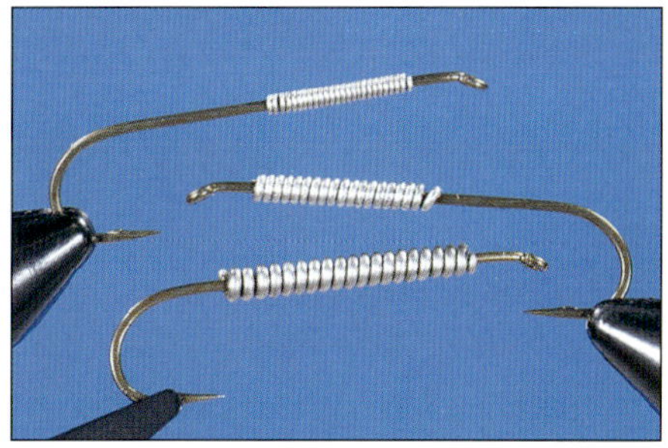

The longer the shank, the more weight you can apply. The three 6X long Tiemco 300 hooks are weighted light, medium, and heavy.

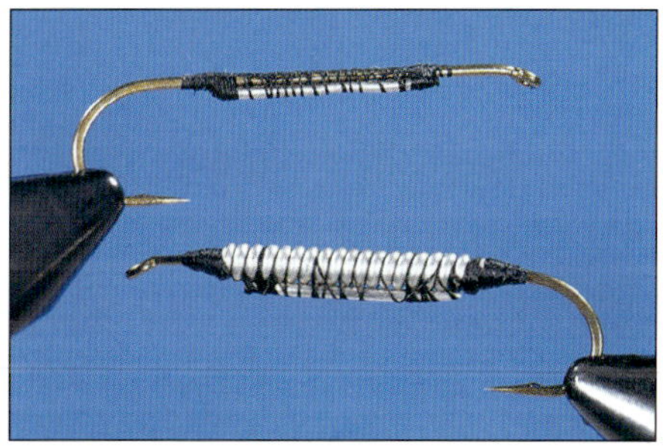

A section of wire tied along the underside of the hook helps it ride keel down. Note that thread has tapered the ends of wire. Secure wire with thread and maybe Zap-A-Gap.

Chapter 5

Weighting Flies

Wire, Beads, and Eyes

When you buy a fly, you do not usually have a choice about its weight or sink rate. If its weight suits your needs, fine—but suppose it doesn't? Perhaps you want to fish the fly in the surface film, or over a weed bed that is eight feet below the surface, or three feet deep in a fast current. One weight cannot cover these varied possibilities.

When you tie your own flies, you can tailor the weight to your specific needs. This is a tremendous advantage. It is a good idea to tie the same pattern in two or more weights. If you cannot tell them apart by looking at them or bouncing them in your hand, change the thread color or make some other subtle change.

Weight is added to hooks by several methods. Wrapping wire (lead or tin) around the hook before tying the fly is the most common. Lead is a major source of toxic pollution and is illegal in national park waters. I suggest using nontoxic, lead-free tin wire. Tin weighs less but the small amount that goes onto a size 14 nymph is of minor consequence. One foot of .025 diameter lead weighs about 1.1 grams compared to 0.75 for tin. See the charts for comparison of bead and wire weights.

You should know the various wire diameters and which is best for a particular hook size. For instance, .055 diameter wire does not fit on a size 16 hook. A few turns of .010 diameter wire does not add much weight to a size 2 hook. Become familiar by wrapping various diameter wires onto different sizes and lengths of hooks, evaluating them for weight and for finished proportions. The larger the diameter of the wire, the heavier and fatter the fly. If you are tying a damselfly and wish it to be lightly weighted, you would use a fine diameter wire. Basically, this is common sense.

Wire is best wrapped over a thread base. If you are weighting hooks in advance of tying, forego the thread base. Wire should usually be placed on the forward area of the hook shank. This allows the fly to ride on a reasonably even keel and leaves room behind the wire to work a taper into the abdomen. Don't crowd the head area. Leave space to taper down the thorax area and to tie off all materials. The top photograph on this page shows how to vary weight using wire. The extra heavy fly may be somewhat bulky, but this can be a worthwhile trade off, especially when you need to sink deep quickly in fast water.

To remove the leftover wire ends, place a finger from one hand on the wrapped wire, grasp the loose end of the wire close to the hook shank with your other hand, and *pull or twist* the wire apart. Tuck any loose ends down. I don't like to cut it because the resulting sharp edge

Wire Diam.	Actual Size	Approx. Hook Size
.010	———	to 22
.015	———	to 20
.020	———	to 18
.025	———	to 16
.030	———	to 14
.035	———	to 12
.045	———	to 8
.055	———	to 4

Wire Weight Comparison

Diameter	Lead Grams/Ft.	Tin Grams/Ft.
.015	.43	.27
.025	1.12	.75
.035	2.18	1.43

can cut your thread, and wire is hard on scissors. If you do cut it, do not use your best scissors.

Secure wire in place by wrapping the thread forward and backward until the wire does not spin around the hook shank. If you wish to create a flattened body, crimp the wire flat with *smooth-nosed* pliers *after* it has been secured in place. If you crimp it too tightly, it will crumble. A drop of Zap-A-Gap helps to ensure that the wire stays in place.

A heavily-weighted fly, especially one tied on a long-shank hook, tends to ride upside down in the water. A strip of wire tied along the underside of the hook helps keep the hook point riding down so fish become hooked in their lower mouths. A wide-gape hook in an upside-down position may be wide enough to puncture the fish's eye or brain when lodged in the upper mouth.

Another method of adding weight is with metal beads, coneheads, or hourglass eyes. Beads are slipped over the back of the hook and materials hold it against the hook eye, while eyes are secured onto the hook with an X wrap. Additional thread wraps should be taken around the base of the eyes to prevent them from rotating. I also like to add a drop of Zap-A-Gap. Paint the eyes any color you like, but do so before you tie the fly. An overcoat of five-minute epoxy increases the durability of the paint.

A 3/32-inch-diameter tungsten bead incorporated into a size 18 nymph greatly enhances its sink rate. It can be difficult to get small flies to sink quickly, but this method allows anglers to fish small flies much deeper without creating bulk. Briefly, beads are available in copper, brass, gold, black, or nickel-silver finish, and several painted colors and sizes. Besides the weight factor, beads and eyes help create realism and attention-getting flash. Glass beads and burned monofilament eyes are used solely for effect. They can create flash and translucence and represent various aspects of an insect.

To create a flattened appearance, smash wire using a smooth-nosed pliers.

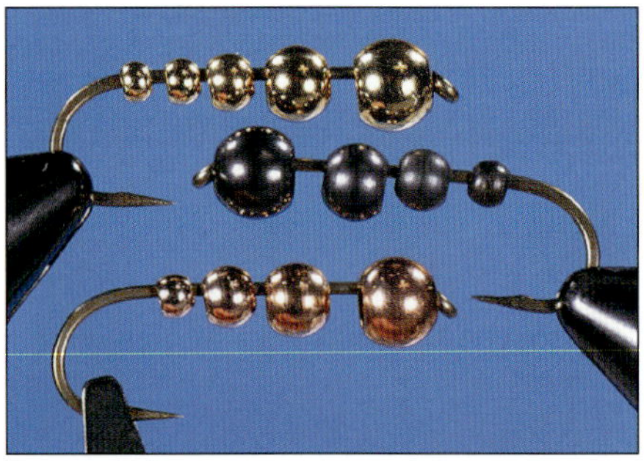
Beads are available in many colors and sizes.

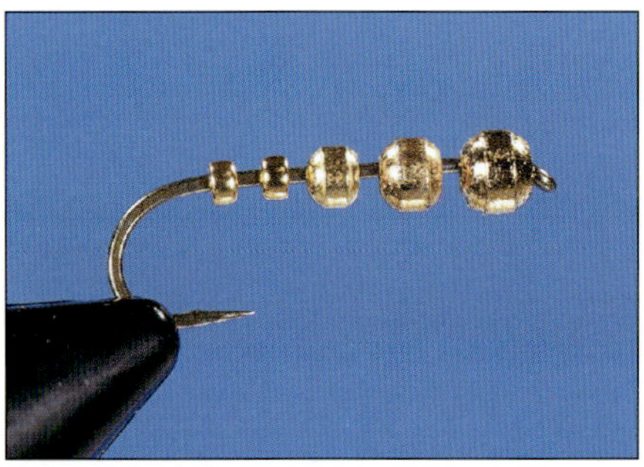
Tungsten beads are the heaviest, two to three times heavier than metal.

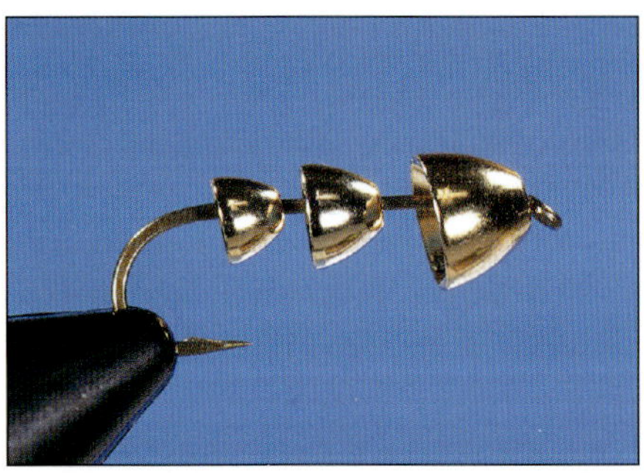
Coneheads add considerable weight and flash and are especially useful on larger flies.

Tin and lead wire is spooled and available in several sizes.

Bead Comparisons
size and weight (in grams)

	5/64"	3/32"	7/64"	1/8"	5/32"	3/16"	1/4"	3/8"
Tungsten	.05	.07	.16	.25	.38			
Metal		.03		.08	.15	.26		
Brite	.02	.03		.08	.18	.27		
Hot					.18	.28	.49	
Cone					.13	.18	.56	1.73

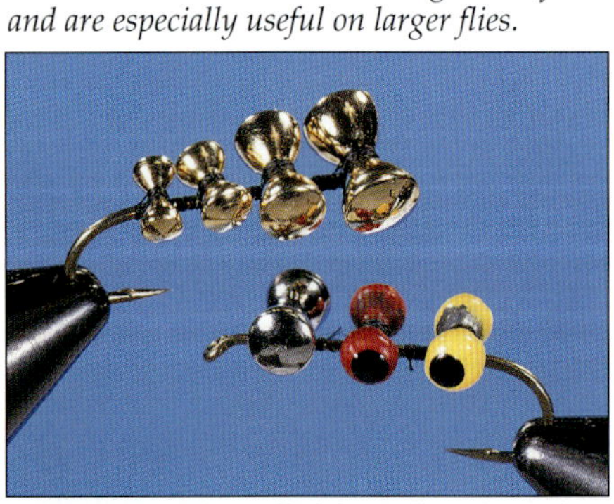
Metal eyes, also called barbell or hourglass eyes, are heavy and available in many colors and sizes.

Glass beads do not add much weight but are of value for effect.

Burned monofilament eyes add realism on some nymphs. Different shapes are sometimes available.

Tying Nymphs

Nymphs, for the most part, are some of the easiest flies to tie and I always urge beginning tyers to start with them. Nymphs can be complicated but many of the most successful flies are quite simple, such as midges, soft hackles, and leeches.

Taxonomically speaking, nymphs are the immature larval stage of aquatic insects that eventually mature and transform into winged insects. Some of the more noted nymphs, or aquatic insects, include mayflies, stoneflies, caddisflies, midges, damselflies, and dragonflies.

Many subaquatic inhabitants exist that do not transform into winged insects and are not classified as nymphs. These are referred to as permanent organisms and include leeches, worms, and scuds. Anglers and tyers sometimes loosely apply the term nymph to the larval and pupal stages of aquatic insects and to permanent organisms. In other words, a leech is not really a nymph because it never transforms into an air breathing or flying stage and always remains a leech.

Nymphs and other subaquatic inhabitants are usually the mainstay of a trout's diet, and anglers who have a basic understanding of nymphs, their habitat, and how, when, and where they are available to fish, increase their hook-ups dramatically. There is some truth to the adage that 10 percent of the anglers catch 90 percent of the fish. Understanding the foods of trout, whether they are crustaceans, beetles, smelt, or eggs is an important aspect of the game. Knowledge translates into success. Tyers interested in learning more about imitating, tying, and fishing nymphs, from the simple to the complex, should consider my book, *Tying Nymphs*.

The following tying instructions illustrate in explicit detail many of the techniques required to construct nymph, streamer, steelhead, and dry flies. The 28 demonstration flies have been selected for their tying techniques, use of various materials, and for their effectiveness and angler interest. They encompass many major food sources, including mayflies, caddisflies, midges, damselflies, leech, and forage fish. Tying history and angling strategy are also discussed.

Flies and tying techniques follow a logical sequence, each suceeding pattern building upon the previous one. You should progress from start to finish without deviation. The photos are exact. Pay careful attention and your tying skills and enjoyment level should increase dramatically. Once you have studied and mastered all the demonstration flies, you should be able to "read" many nymph, dry, streamer, and steelhead patterns. A fly "pattern" is a parts and material list that explains how the fly is tied and helps you visualize what it should look like.

Nymph Proportions (5262)

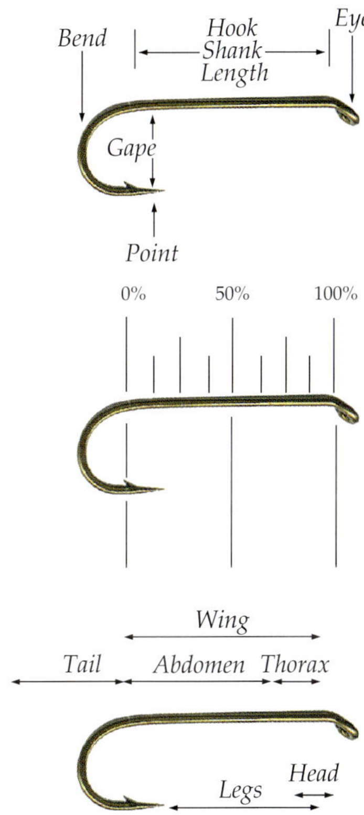

Classic nymph "parts" and proportions. These are general guidelines that vary according to tying and fishing style, the intended food source being represented, hook style, and fly pattern.
Tails are generally 1-1½ times as long as the hook gape is wide.
Abdomen usually occupies the hook shank to the 95 percent mark. If there is both an abdomen and thorax or head, the thorax begins at the 60-70 percent mark and continues to the thread head location.
Heads begin at the 80-85 percent mark. Thread heads occupy the front three to five percent.
Legs usually extend to the hook point or less.
Wings extend to the end of abdomen or less.

Always smash barbs when hooks are not barbless.

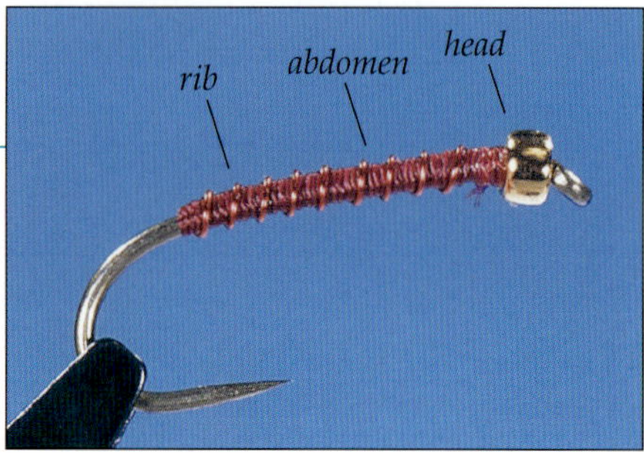

Bead-Head Midge
Hook: 3761BL, 900BL, or 2487BL, sizes 12-18
Head: Copper bead
Thread: Red
Rib: Copper wire
Abdomen: Red thread
Note: Other color combinations to suit.

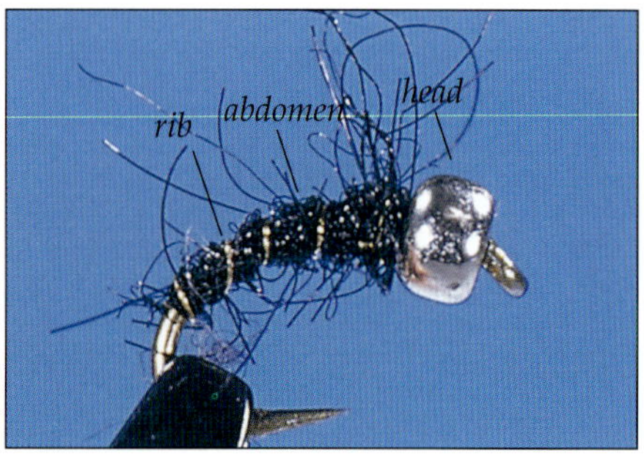

Micro Bead-Head Midge
Hook: 2487BL, or 900BL, sizes 18-22
Head: Tungsten black/nickel bead
Thread: To match abdomen
Rib: Wire, your choice
Abdomen: Ice Dub, Antron or any other dubbing

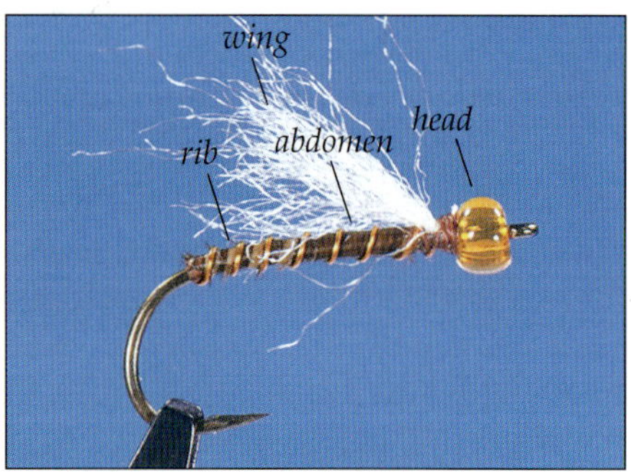

Glass Bead Emerging Midge
Hook: 101 or 900BL, sizes 12-20
Head: Amber glass bead
Thread: Brown
Rib: Copper wire
Abdomen: Brown turkey biot
Wing: Clear (white) Antron yarn
Note: Other colors to suit

Chapter 6

Midges

Bead-Head, Micro Bead-Head, and Glass Bead Emerging Midge

Midges belong to the Order Diptera and are the most prevalent insects in the world of trout. They span the color spectrum, and can populate to thousands of individuals per square yard of bottom strata. Midges are friendly, non-biting insects, and anglers often observe clouds of swarming adults about them. Adult midges can be differentiated from adult mosquitoes because midges do not have a biting proboscis or raise their hind legs when at rest like mosquitoes do.

Midge larvae and pupae provide many meals for trout, and it is not uncommon for lake anglers to hook fish of several pounds on size 16 to 20 subsurface midge imitations! Some midge species are represented by hook sizes 8-12, but these are the exception. Most are thinner than the hook is thick, so it is almost impossible to tie them too skinny. When you weight the hook with wire, they become too fat. The best way to construct weighted imitations is with a tiny tungsten bead.

Within the past 10 years, beads have been incorporated into almost every subsurface pattern. Beads are manufactured from metals and glass. Glass beads weigh little and are used for effect—light reflection, translucence, and profile. They are available in many colors and sizes, and have a straight drilled hole.

Metal beads are usually manufactured from brass and tungsten. They are available in many colors, and are used for a combination of weight and flash. Most have a tapered drilled hole, making them easier to slip around the hook bend and snug against or behind the hook eye.

As you stock your fly boxes, think where and how you may be fishing. Consider the fly line, leader length, hatch, and stage of the insect you should imitate. This dictates the size, color, sink rate, etc. If you are fishing midge larvae in a lake 12 feet deep, slide a tungsten bead onto the fly. For a surface film "emerger" presentation, tie an emergent wing on a weightless or slightly weighted midge, etc.

How many midges do you need? That depends on where, how, and how long you fish. If you stock only three colors with two different colored beads or no beads in three sink rates—you do the math while you keep tying!

The following midges demonstrate how simple fly tying can be and that flies need not be elaborate. The following three flies are a study in beads, sink rate, forming an abdomen (body), installing rib, and an emergent "wing."

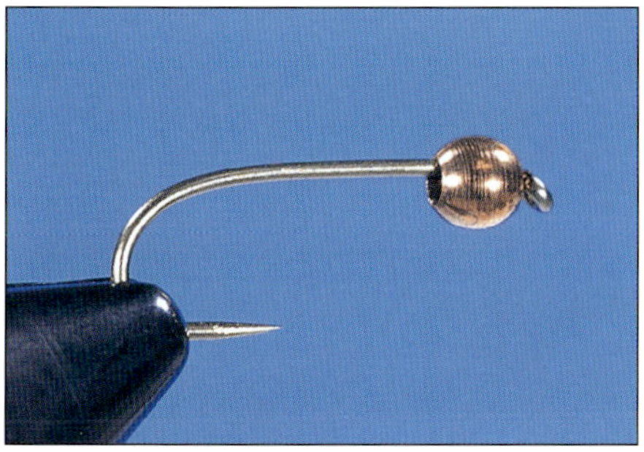

Bead-Head Midge
1. Insert the smaller tapered end of the bead over the hook point and around the hook.

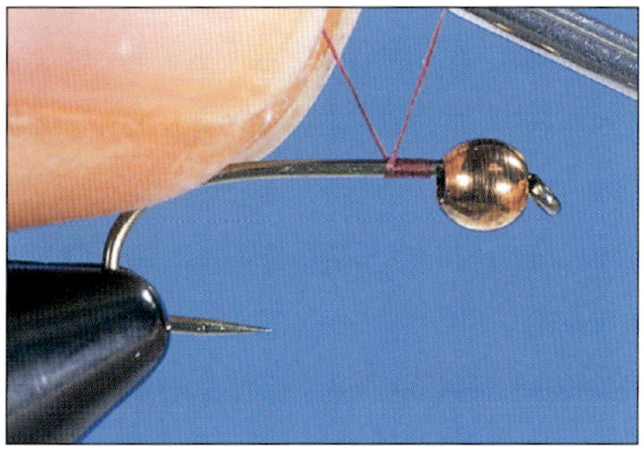

2. Secure the thread at the position pictured, trim off the thread end and wrap the thread back toward the bend of the hook. Thread eventually forms the body.

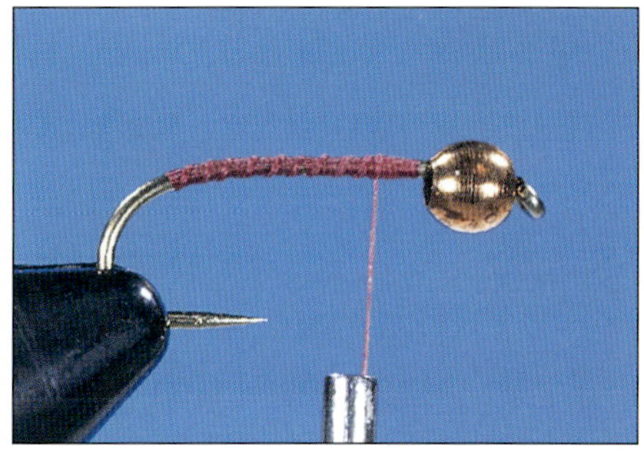

3. Position the thread as pictured. The wire, which forms the rib, is tied in at this location. Tie in the wire, using the pinch or cinch technique, along the top of the hook shank.

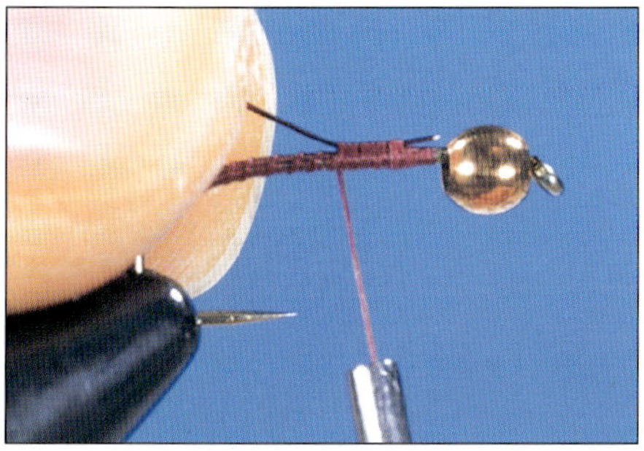

4. Hold the wire above the hook as pictured when you tie it down along the top of the hook. This ensures that it stays in place directly and evenly on top of the hook so it can't roll to one side, causing an uneven foundation.

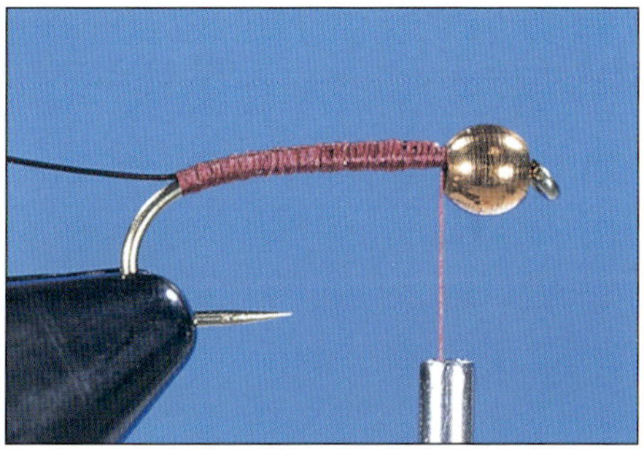

5. The wire has been secured back to the hook bend (end of abdomen) and the thread is repositioned behind the bead, forming abdomen. If you placed the thread wraps tightly side-by-side, the abdomen (body) should look like this.

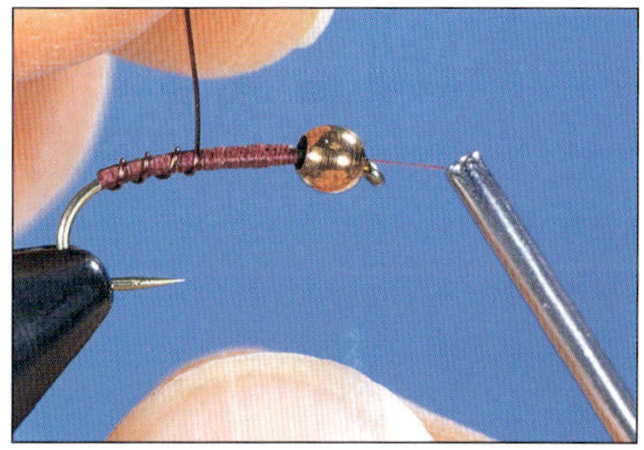

6. Using the over-the-top and hold-it technique, wrap the wire tightly and space evenly through the abdomen. Note the position of the hands and bobbin, which rests over the right hand out of the way.

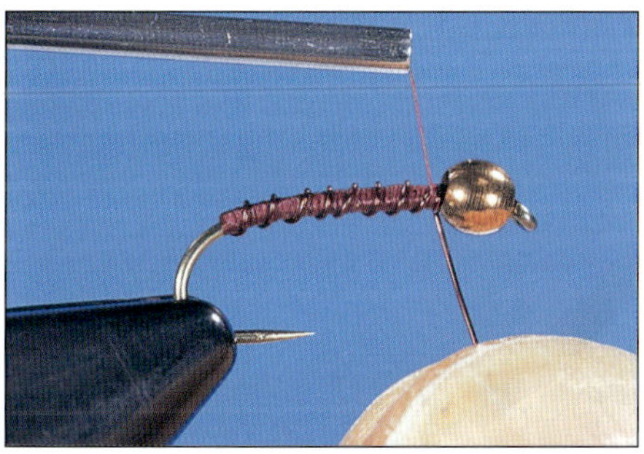

7. Take an extra turn of wire behind the bead and tie off the wire. This extra turn makes it easy to secure as the thread wraps over an entire turn of wire instead of only the wire end.

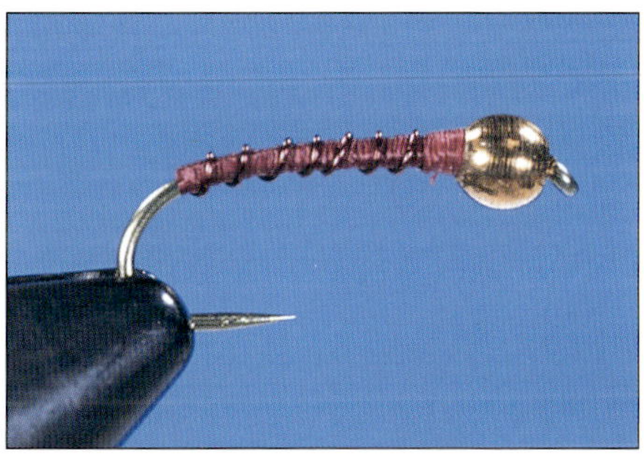

8. Tie off the thread head behind the bead. Finished Bead-Head Midge, which represents larvae. Actual midge larvae are usually tiny worm-like creatures that span the color spectrum. Black, olive, and maroon-red are favorites with anglers.

Simple flies can be very effective. Dubbing or other material can also be used to create the midge abdomen, which is demonstrated next.

Let's tie a small fly. The Micro Bead-Head Midge could be tied with only a thread abdomen but practice dubbing the thinnest possible amount of Antron.. Remember, small flies require few wraps and little material. It is almost impossible to put too little material on small hooks.

If you tied the Bead-Head Midge without difficulty and practiced dubbing in the earlier chapter, the Emerging Midge should also be easy. If you have trouble, refer to page 28.

Micro Bead-Head Midge
1. Actual hook pictured is a 2457, size 18. It does not require much material to cover this size hook. Slip a tungsten bead onto the hook, and wrap thread over the hook as shown.

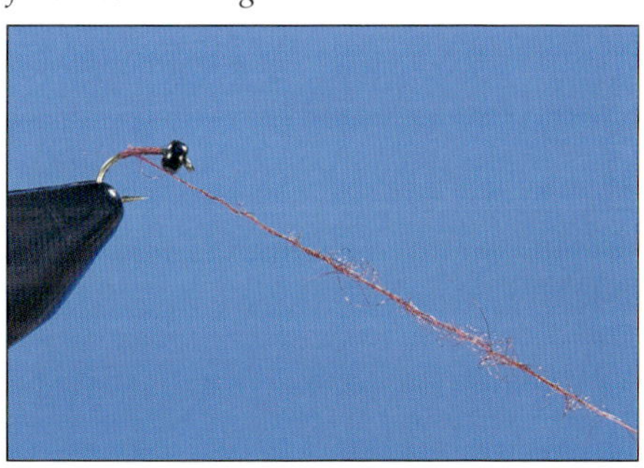

2. Pull the bobbin thread down about three inches from the hook and apply dubbing. Use smooth Antron; it is easy to dub. Use very little. You need only change the color and texture of the thread.

3. Wrap the dubbing around the hook, beginning at the back and ending at the front. Add a little more to complete a taper if needed, or pull off any excess. Tie off the thread head—that's it! Too easy. Try the next one!

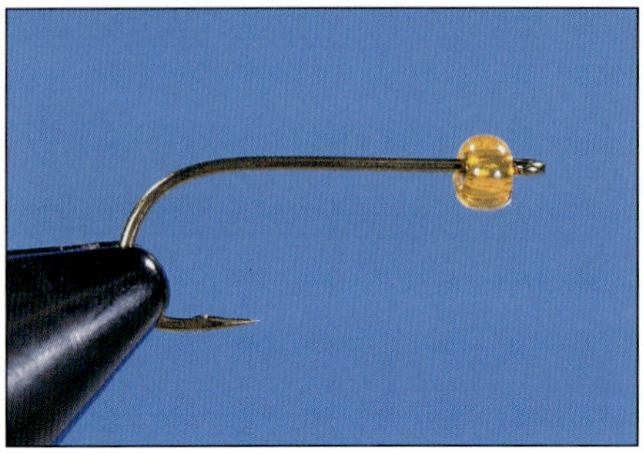

Glass Bead Emerging Midge

1. The 101 hook has a straight eye. This is sometimes preferred for midge, especially small sizes, because the straight eye does not protrude into the hook gape like a down-eye might.

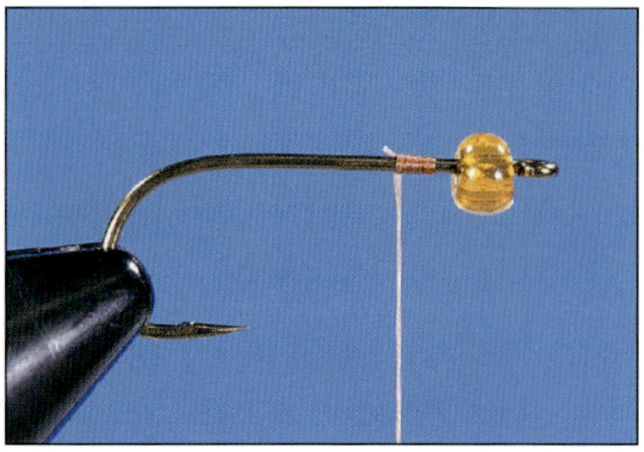

2. Glass beads are usually drilled with a straight hole. Be certain the hole is small enough so the bead can snug against the hook eye but does not slide over it. Use any color or weight bead you like.

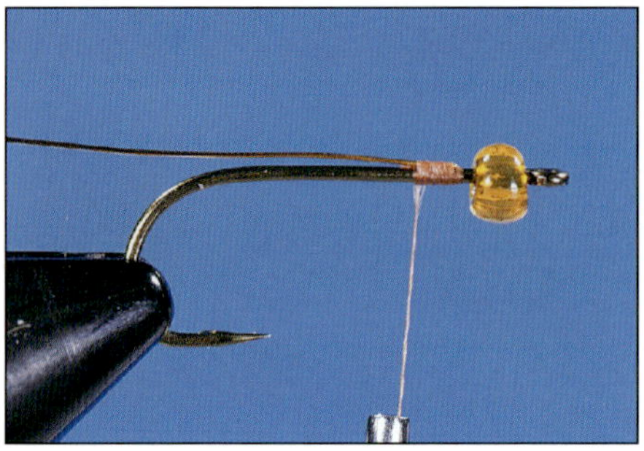

3. Tie in a section of wire. The diameter is determined by the fly size and desired results. We are using size brassie (the name for a particular size of wire) on this size 10 hook.

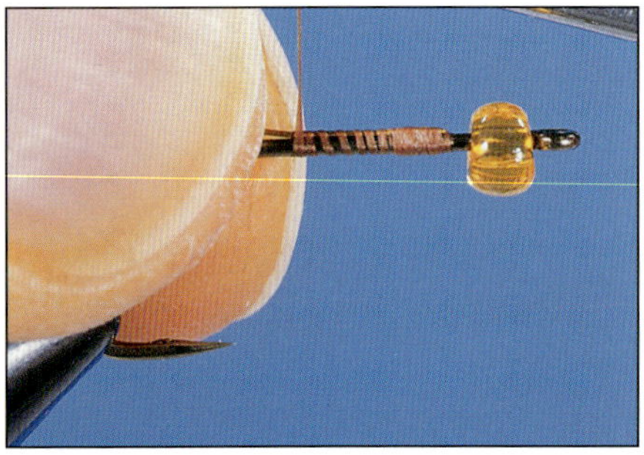

4. Remember to secure the wire directly along the top of the hook, keeping a level foundation. This technique holds true for any and all materials that are installed. Keep the underbody even.

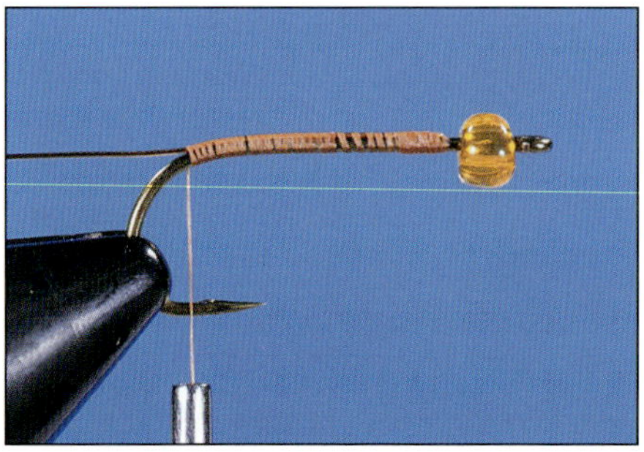

5. Perfectly installed wire and even (level) underbody.

6. If your underbody looks like this, try it again.

7. Select an individual biot quill from the short side of a turkey biot (quill) feather. Biots are used for nymph abdomens, tails and antennae (Copper John, Prince) and for spent spinner and adult dun bodies.

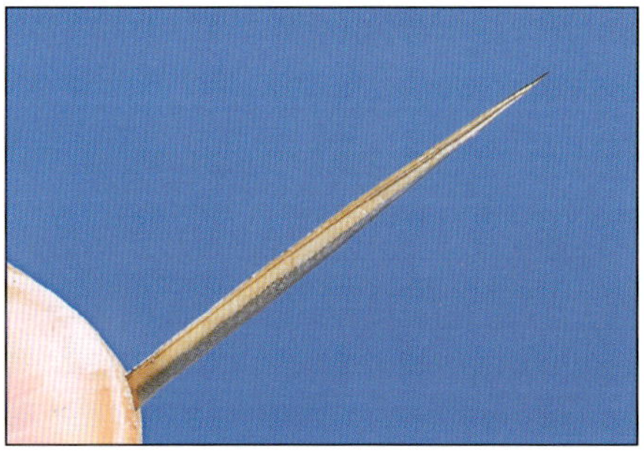

8. A single biot quill. Because they are short, they are best handled with hackle pliers. There is a ridge or raised edge along one side of the feather that, depending upon how it is wrapped, creates a natural ridged ribbing or segmented effect.

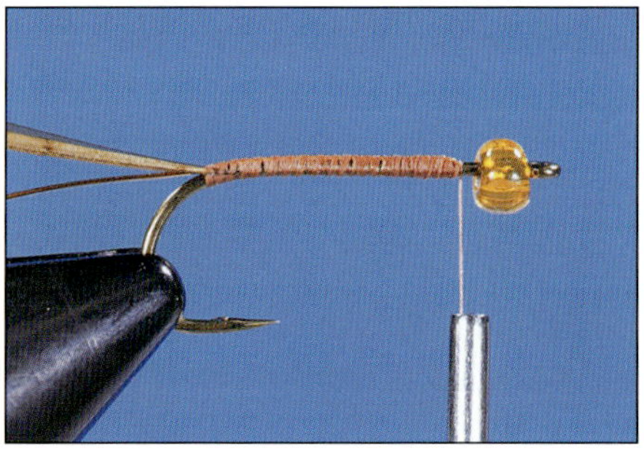

9. Tie in the biot quill on the opposite side of the hook by the extreme tip. Do not disrupt the level foundation. Wrap the thread to the tie-off point as pictured. Biot feather forms the abdomen.

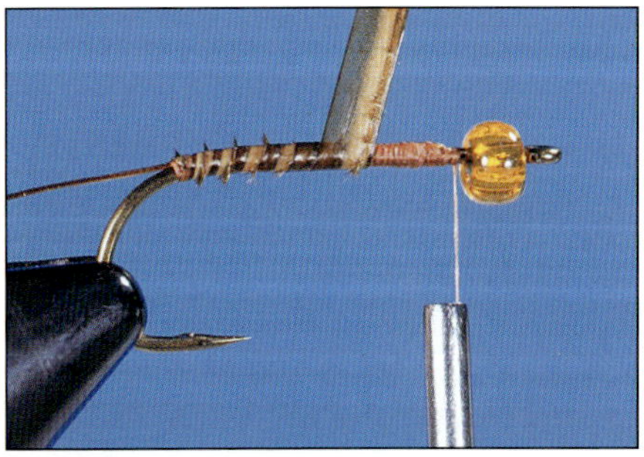

10. Use hackle pliers to wrap the biot quill. The ridge effect disappears at the 50-percent mark because the width of the feather and angle of the wrap cause an overwrap of the feather, covering the raised edge.

11. Tie off biot quill behind bead. Wrap rib. Notice position of hands and bobbin. Wrap wire over the top of the hook with left hand. Hold it at bottom with right hand. Pick up wire with left hand and repeat.

12. Take an extra turn of wire, then tie it down by holding wire in right hand, bringing thread around, behind, and over wire. Keep the bobbin close to the hook, working a small circle of thread.

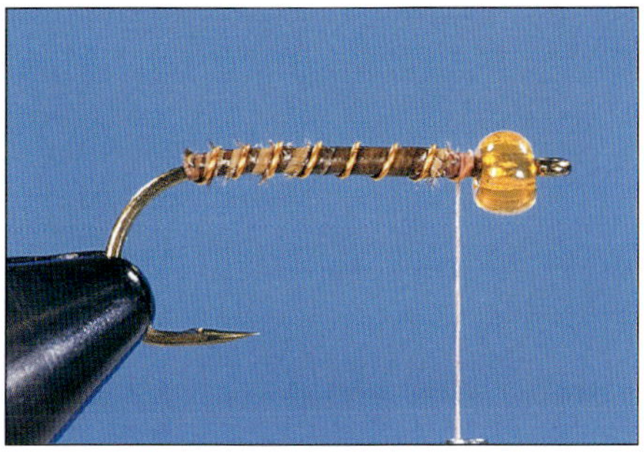

13. The Glass Bead Midge is complete—or we can continue and tie on a wing, creating an Emerging Midge. This represents the final underwater stage as the larva prepares to transform into an adult.

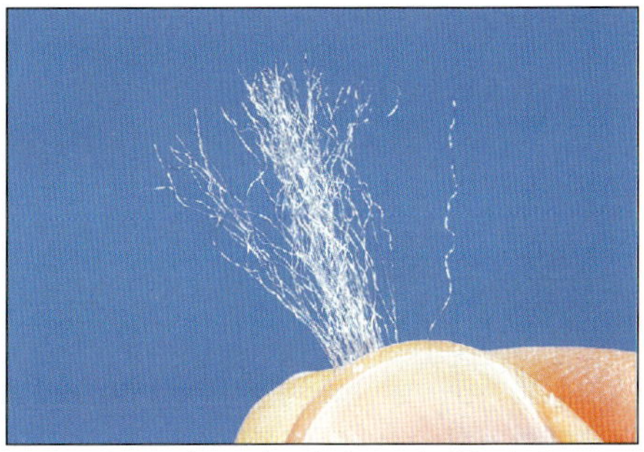

14. Trim a length of clear (white) Antron yarn. Antron offers a loose, or open, translucent effect, and it is easy to work with and trim to shape.

15. Thread should be positioned slightly behind bead (photo 13). Hold fibers in right hand and place over hook. This is about the right length, but this material can be trimmed to length and shape so that perfect placement is not necessary.

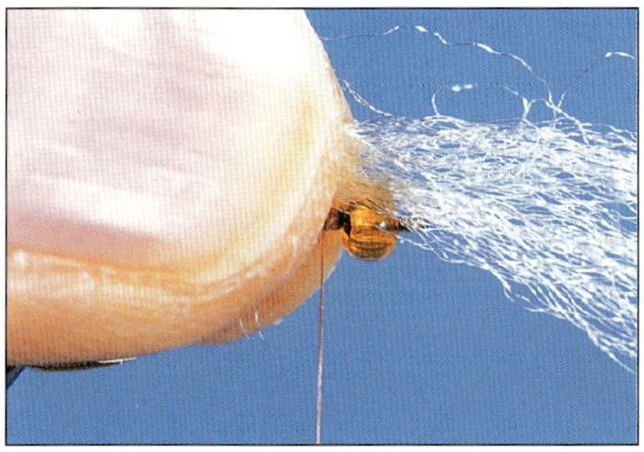

16. Change hands, grasping wing with left thumb and first finger, pinching thread between fingers. This holds the wing in place as you secure it with thread at this exact position.

17. Note position of thread slightly behind bead. Use the up-between-the-finger pinch technique to secure, or try a loose wrap over the top and cinch downward, securing wing exactly over top of abdomen.

18. Wing is tied in place.

19. Position thumb and first two fingers so they encompass the hook eye. With fingers, sweep all fibers back away from the bead. Take a few wraps of thread at the position pictured, keeping all fibers clear of bead and head area.

20. There are no errant fibers; fly is almost complete. Follow the pictured instructions and make a close cut of the leftover wing butts.

21. Grasp all fibers with right hand. Using a sharp, fine-pointed, thin scissors, place tips tightly against the hook and eye as pictured and make a close cut. If you miss some, cut again.

22. Wrap a few more turns of thread to cover any stubble, half hitch twice, cinch half hitches backwards until tight, lacquer, and that is it! If necessary, trim the wing to length and shape. Nothing to it!

23. Midges come in many colors and sizes. Consider the weight factor and the stage to imitate as you stock your fly box. Notice the raised segmented effect of the biot quill on the size 18 yellow midge.

24. The olive size 10 "magnum" midge contrasts sharply with the size 18 micro midge. The olive midge has a dubbed thorax and could also imitate caddis larvae. Continue on to the next chapter!

Rhyacophila *Caddis*
Hook: *Tiemco 2457, 12-16, weight to suit*
Head: *Bead (optional); color, size, weight to suit*
Thread: *Black*
Rib: *Wire, thread, or floss twisted tightly; color to suit*
Abdomen: *Creamy mint green to bright metallic green or hare's ear dubbing; consider HareTron, SLF, or squirrel*
Thorax: *Peacock or slight amount of black HareTron blended with body material, picked out*

Rhyacophila *Caddis Emerger*
Hook: *2457, 12-16, weight to suit*
Head: *Bead (optional); color, size, weight to suit*
Thread: *Black*
Rib: *Wire, thread, or floss twisted tightly; color to suit*
Abdomen: *Creamy mint green to bright metallic green or hare's ear dubbing; consider HareTron, SLF, or squirrel*
Thorax: *Peacock or slight amount of black HareTron blended with body material, picked out*
Legs: *Brown partridge fibers tied divided style along each side of abdomen*

Note: Legs can extend from behind thorax, in front of thorax, or from near the front of thorax. When legs are tied at front of thorax, a turn or two of dubbing covers the tie downs.

Chapter 7

Rhyacophila *Caddis and Caddis Emerger*

Caddisflies belong to the Order Tricoptera and are found in almost all trout waters. There are over 900 species in North America, and they come in many sizes and colors. In sheer numbers, they are second only to midge.

Larvae look like worms. Some build cases, nets, nests, or live in cocoon-like structures. Others are free swimmers and seek shelter among rocks in quick-flowing water. The free swimmers are the easiest to imitate and fish. *Rhyacophila* is the best known of the free-living caddis in the West. They are commonly found in sizes 14-16 in olive, bright green, blue-green, and shades of tan. To imitate other caddis larvae, simply change the color, size, and weight to fit your needs—the shape remains basically the same. Any dubbing material works, natural or synthetic.

When larvae (or pupae) are "in the drift," they can be stretched out straight in a swimming position, curled into a ball for protection, or slightly curled. The slightly curved shape is the current favorite of tyers and anglers. There are so many identical "patterns" of caddis larvae and pupae that they have become a tying style. Larva patterns are easily converted into pupa patterns by adding emerging wings, which can be tied with many materials, including partridge, duck, CDC, Antron fibers, or deer.

Experimentation and creativity are important and fun aspects of tying your own flies. Remember, no patterns are cast in kryptonite. Tie with materials you like and design flies for yourself.

The *Rhyacophila* Caddis Larva is a practice in weighting, dubbing, ribbing, and constructing emergent wings/legs.

When done properly, dubbing is simple and easy. Spread dubbing material apart so it is "loose." Then, spread a thin amount evenly along a two-inch section of thread, beginning about one inch down the thread from the hook point. Remember to just barely cover the thread. Keep the bobbin/thread taut with one hand and roll the thread and dubbing in one direction under pressure between your thumb and first two fingers of your other hand.

The thread and dubbing should turn or roll with or between your fingers. Do not roll the thread and dubbing back and forth as this only reverses what you did. Roll only in one direction, exerting lots of pressure between your fingers. Pretend you are melding the dubbing and thread together. Begin at the top-most dubbing fibers along the thread and work your way down the thread away from the hook. One or two rolls should do it.

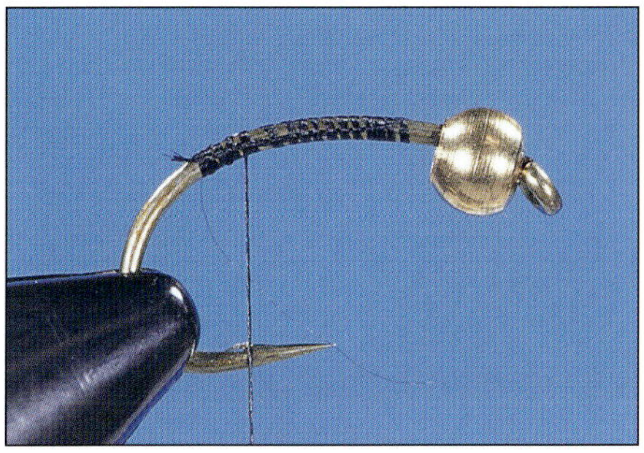

Rhyacophila *Caddis*

1. Insert bead on hook, cover with thread. Wrap wire for added weight. This is placed in the middle area and somewhat forward to facilitate a tapered abdomen, leaving space behind bead for other materials to come.

2. Secure the steel wire by wrapping thread over and through the wire towards the bead, then back over wire. Tie in wire rib and prepare a thin strand of dubbing. Test the wire—it should not rotate easily around the hook shank. Fine-diameter wire is easy to secure.

3. Wrap dubbing to just short of bead as pictured, keeping abdomen even and slightly tapered.

4. Wrap the wire rib forward and secure behind bead. Rib is usually intended to create subtle flash and segmentation and to reinforce the material it is wrapped over.

5. Dub a sparse amount of black dubbing, creating the thorax, which is usually slightly larger in diameter than abdomen. This finishes the Rhyacophila, *but let's continue and construct an emergent wing.*

Rhyacophila *Caddis Emerger*

6. The partridge feather, as we are installing it, can represent legs on nymphs or, in this case, the emerging-expanding wings/legs of the caddis pupa. Select a brown partridge body feather that is well barred.

7. Trim out the center stem and pull off the fuzzy and shorter fibers toward the base of the feather as pictured. The center stem controls too many fibers and is difficult to divide, so eliminate it.

8. Hold fibers in right hand, splitting them evenly along each side of the fly. Tips should extend about to the hook point. When the length is correct, pinch them in place with your left hand.

9. Pinch in place, then secure in place along each side of the fly. Take a loose turn with thread and tighten fibers in place. Your fingers control the position or where you allow fibers to be tied down.

10. Top view of wings/legs. Many patterns are tied with the fibers (legs) secured in place only along the underside of the fly. These are usually referred to as "beard hackle." Trim ends closely behind bead and tie off head.

11. If you cannot trim fibers closely enough and leftover ends show, dub a little more black HareTron dubbing as pictured and finish the Rhyacophila *Caddis Pupa.*

12. This pattern can represent most species of caddis larvae. Tie on a curved or straight hook in appropriate colors, sizes, and sink rates. As pictured, these could also represent scud.

Marabou Blood Worm
Hook: 3761BL, sizes 10-16, or 5263BL, sizes 10-14
Head: Gold bead
Thread: Maroon
Rib: Copper wire
Tail: Maroon marabou
Abdomen: Maroon marabou

Marabou Leech *(Hal Janssen, 1961)*
Hook: 5263BL, sizes 4-10
Thread: Black, or color to match body
Tail: Marabou, color of choice with olive Krystal Flash highlights
Wing: Marabou, two to three separate wings staggered along hook. Highlight with red, gold, black, peacock green Krystal Flash
Head: Marabou
Note: Marabou wing is tied in three or four stages and actually represents the leech body. Each bunch is the same length (but tied in at a different location so the marabou is staggered and does not extend to the same point). This makes the wings "stack" and widest or most dense at the center point and tapered at the ends, much like a leech.

Marabou Damsel *(Randall Kaufmann)*
Hook: 200RBL, sizes 8-12
Thread: Olive
Eyes: Burned monofilament
Tail: Olive marabou
Rib: Gold, green, or copper wire
Abdomen: Olive Marabou
Wing: Olive marabou and olive Krystal Flash

Chapter 8

Marabou Techniques

Marabou Blood Worm, Leech, and Damsel

Most marabou used for fly tying is harvested from immature turkeys. It is packaged under mini, Woolly Bugger, and choice grades, and is two to six inches long and one to four inches wide. Select smaller marabou for smaller flies. A thin center stem runs the length of the feather and has about 40 flimsy fibers per inch extending from two sides. Each fiber has dozens, perhaps hundreds, of minute barbules extending up to one-sixteenth-inch from two sides of the fiber. It is very animated, both in and out of the water.

Marabou is one of the few materials that can be altered without sacrificing too much of its natural appearance. Break or pull it apart with your fingers—it still looks and acts natural. Trimming it with scissors is a last option, but marabou remains relatively unchanged.

This versatile material is encountered in tying nearly all styles of flies, including eight of our demonstration patterns. The following three flies depict how to handle and tie marabou as a tail, wing, and abdomen. The Marabou Blood Worm is tied first. By altering the color, weight and size of this simple pattern, it can imitate leech, damsel, and possibly scud.

The Marabou Leech is next and is also easy to tie. The endless color combinations stimulate the imagination. A fly box full of these jolts the senses, and they fish well as leech, minnows, and attractors. Maroon, red, white, orange, and purple have goaded many oversized trout into attacking my offerings over the years.

The Marabou Leech was developed by expert California angler/tyer Hal Janssen in 1961, before the advent of flashy materials, and was originally tied in solid colors. Try two, three, or four different color wings and add Krystal Flash. I hope Hal doesn't mind my messing with his pattern! Constructing the "wing" is the same principle for tying many standard streamer wings, and many forage fish color combinations can be imitated with this pattern. More on streamers in chapter 13.

Last is the Marabou Damsel. It is a bit more tricky and deals with one-piece construction and monofilament eyes. Damsels are magical insects belonging to the Order Odonata. They are beautiful, friendly, animated creatures that inhabit waters rich in aquatic vegetation. Damsels and their cousins, dragonflies, are especially plentiful in lakes, where they provide some of the most unique and exciting trout fishing to be found. Every angler in-the-know anxiously awaits the first damsel migration of the season.

Marabou Blood Worm, Method 1

1. Insert bead and cover hook with thread. Position thread at bead. Remove a bunch of marabou. Align at right angle from stem and pull off. Fibers have been dampened for ease in handling.

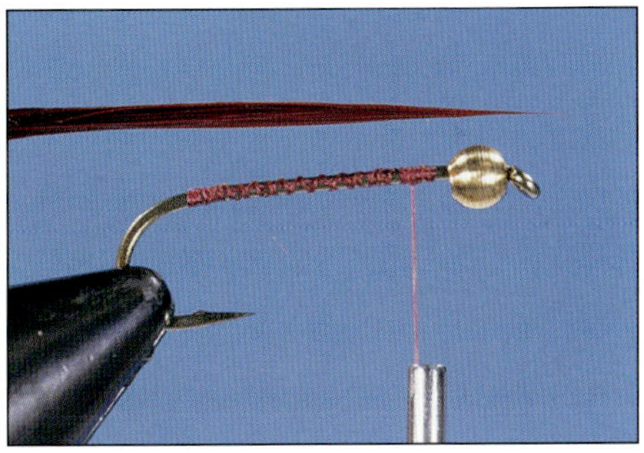

2. The fibers are tied in tip first. This photo merely demonstrates that fact. Reposition fingers at marabou tips and tie in place behind bead using up-between-the fingers pinch technique.

3. Marabou has been tied down along the top of the hook to the tail area. The marabou tips need to be cut at the bead (next photo) and a length of wire tied in at the tail area for rib.

4. Best method to trim material close is to grasp all material to be cut and place tip of scissors against hook, thereby insuring the closest possible cut. Tie in copper wire for rib. It is behind the marabou and difficult to see.

5. Half of the marabou tail has been used to wrap the abdomen. This one-piece construction is similar to the Marabou Damsel, which is shown on page 41. The next photo series of the Marabou Blood Worm on this page shows two-piece marabou construction.

6. Finished Blood Worm. Is that easy, or what? Try the next method, where two lengths of marabou are used. This method allows for better control of abdomen taper and a more natural tapered tail.

Marabou Blood Worm, Method 2

1. After inserting bead and covering hook with thread, tie in a bunch of marabou as shown, front to back.

2. Trim off excess behind bead.

3. Select another small bunch of marabou and tie in tip first at tail area. Then, tie in a length of wire for the rib.

4. If the marabou tie in does not alter the taper, trim it off here. Otherwise, tie it down over top of hook to the bead. Then, twist the marabou slightly, wrapping it to form the abdomen.

5. Finish a thread head, and that is it. When marabou is tied in and wrapped tip first, a natural taper develops because the barbules become slightly wider toward the butt of the feathers. Shorten tail as desired.

6. Blood Worms can be tied in any color and sink rate, usually sizes 10-16. The next pattern, the Marabou Leech, is a cross between a nymph, leech and streamer. Try it!

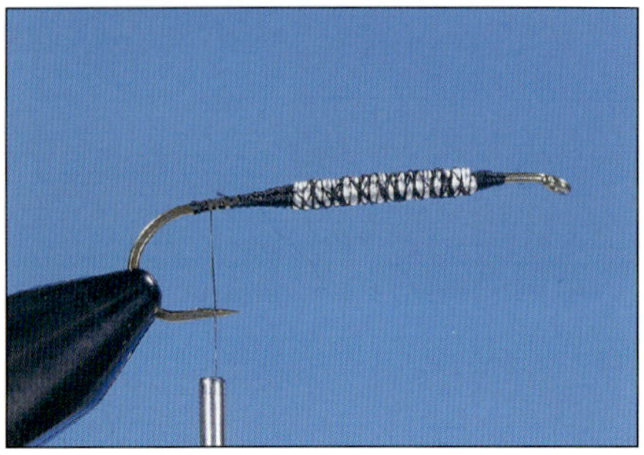

Marabou Leech

1. Weight hook with wire. There is space in front of and behind wire to work a taper into fly and to keep the abdomen level and free of lumps. Four sections of marabou are needed for this length hook.

2. We are selecting four colors of marabou, graduating from light to dark. Use thick, heavy plumes. Fly is constructed so it is heaviest or most dense in its center, helping to create the illusion of a leech.

3. Dampen the marabou for better control. Trim out the tips which rids the plume of the thinnest fibers and the center stem that controls them.

4. After the tips are removed, trim off the base so the remaining center cut is the thick fibers that, when bunched together, are pretty much of uniform length. Measure the plume length against the hook.

5. Tail length (and all other bunches on this fly) should be the exact same length, which is about as long as the hook shank. Secure behind wire.

6. Tie in a three- to four-inch length of pearl Krystal Flash at its center along the near side of the hook. Tie the other end along the far side of the tail. Doing both at once instead of one at a time saves one cut and one tie in.

7. Position thread at about the 30-percent mark and secure another bunch of marabou. Trim off the butt ends. This "wing" is purposely tied long because we want to trim tips back to "thick" area of plume.

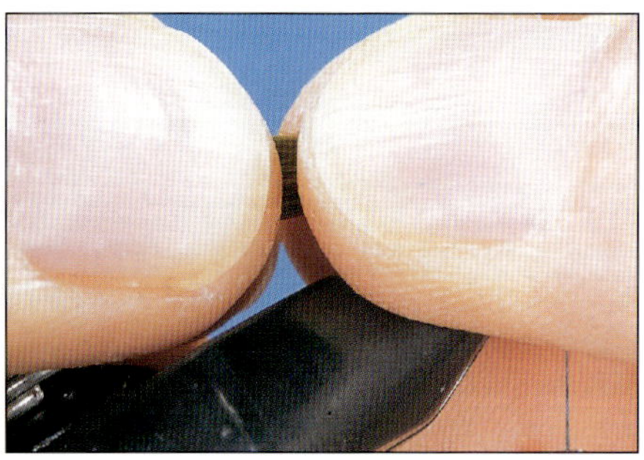

8. This pattern is supposed to be very dense so we shorten the tapered tips to get into the most dense section of fibers, which are closer to stem. Pull marabou apart using your fingers. If you cannot do this, use scissors.

9. Tie in more Krystal Flash along each side of the "wing." You'll notice that this abdomen, or underbody, is only wire and thread. Cover all the wire with thread if you like, or use colored wire. Olive would be best, but black is used for demonstration visibility.

10. Tie in a third bunch of marabou and Krystal Flash as pictured. Shorten tips so this bunch is the same length as others. Bunches are staggered along hook, so the ends do not reach to the same point, but all measure the same length.

11. Secure last bunch, leaving room for a marabou "head." Do not trim ends. We have trimmed out the end stem a half-inch down, leaving fibers still attached and a little stem to hold onto. Wrap marabou around hook, forming "head."

12. Tie off, stroke all fluffy fibers back away from head area and finish thread head and lacquer. Tying these marabou "wings" is the same technique used on streamer wings. More on this later.

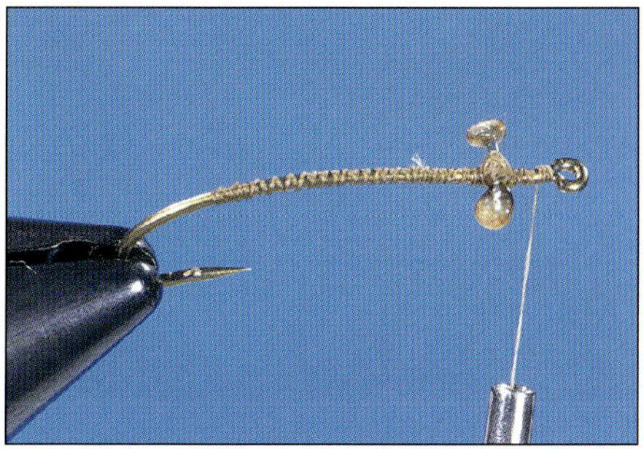

Marabou Damsel

1. Cover the 200RBL hook with thread. Secure the burned monofilament eyes at the 75- to 80-percent mark using an X pattern. A drop of Zap-A-Gap helps secure eyes. Position thread at hook eye.

2. Select a marabou feather (plume) and grasp a clump of fibers, pulling them away from the center stem at a right angle as pictured. This keeps the tips somewhat even. Dampen the feather for better control.

3. The missing fibers from the marabou feather form the tail, body, and wing of this fly. Marabou is one of only a few materials that can be altered with fingers (and, to a lesser degree, scissors) without undue negative effect. Natural tapered ends are usually best.

4. The tapered ends (tips) form the tail (actually gills on a damsel). The butt ends form the wing, and the center section forms the abdomen.

5. Tie in the marabou immediately behind the hook eye and wrap thread back to the front of the mono eyes as pictured. Continue tying the marabou down along the top of the hook back to the tail area. It goes over the top and between the eyes.

6. Keep the marabou centered along the top of the hook, spacing the thread wraps evenly as pictured. This forms the abdomen. Take four to five secure turns at tail area and tie off with two half hitches. Retie thread behind eyes—or, don't cut at tail and spiral forward.

7. Tie in two strands of Krystal Flash that extend the length of the marabou. Position the thread as pictured. The marabou in this photo has been dampened for better control.

8. Double the marabou and Krystal Flash back over the top of and between the mono eyes and secure as shown, forming both the thorax and wing.

9. Finish a knot and trim thread. Notice tail and wing length needs to be adjusted. Trim Krystal Flash with scissors. Shorten tail tips to length with your fingers—they look more natural than if trimmed.

10. If the wing marabou is too tough or short to shorten with your fingers, trim with scissors. If you cut, trim at an angle as shown, and also trim the underside at the reverse angle so the wing lies flat.

11. Finished Marabou Damsel. Note that the tail can be left somewhat long and adjusted once you determine the exact size of naturals when you fish.

12. Damsels are usually shades of olive, tan, and brown and average about ¾ to ⅞ inches in length. These wings have been shortened with fingers for a more natural look.

Soft Hackle, Peacock
Hook: 3761BL, size 10-18
Thread: Black or Olive
Rib: Fine gold or copper wire
Abdomen: Peacock
Legs: Brown partridge, tied full

Soft Hackle
Hook: 3761 or 900BL, size 10-18
Thread: Color to match abdomen
Abdomen: Floss, dubbing, or wire
Thorax: Natural hare's ear or omit
Legs: Brown or gray partridge or mottled brown hen hackle, tied full
Note: Three styles of soft hackles shown.

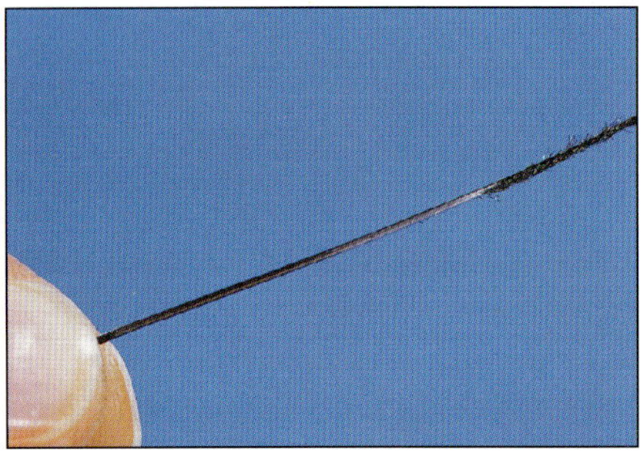

This peacock quill was stripped of herl using an eraser.

Chapter 9

Peacock Techniques

Soft Hackle

Peacock is everyone's favorite material, and trout are attracted to its iridescent metallic properties. Peacock herl is found on eyed tails. The choicest herl is found immediately below the eye for two to four inches. Peacock quills, or herls, are smooth and flat. Herl, the fuzzy fibers, extends from both sides of the quill, but only from one side of the surface area. The herl is longer and most dense on one side. When wrapping herl, the herl-less side of the quill must precede the herled side; otherwise the herl-less portion overwraps and mats down the herl.

Smaller flies require only one herl. Larger flies may require several. Two or more herls must be properly aligned for best results. It is sometimes necessary to form an underbody of dubbing when an extra-juicy effect is desired. Peacock herl is fragile, and care must be taken when handling it. To ensure durability, overwrap or reinforce with thread or fine wire.

Bare or stripped quills should also be overwrapped or cemented. To remove peacock herl, use an eraser or your thumb and first finger. Stripped quills are used on some nymphs but mostly for dry fly bodies, such as the Hendrickson, page 77, and Quill Gordon, page 79.

The soft hackle is believed to be about 500 years old, easily making it the oldest surviving fly pattern. It can represent various stages of many food forms but is usually fished in or just below the surface film. Once again, simplicity is king. The Peacock Soft Hackle is a three-part pattern consisting of rib, abdomen, and legs. The most popular soft hackles have a floss body and dubbed thorax. Practice a peacock body.

1. Begin by selecting a choice peacock tail feather. Select a quill with long, dense herl found immediately below the eye of the tail. Look at a peacock tail feather and notice where the herl begins to diminish in size, ultimately becoming flimsy at bottom of tail.

2. Strung herl includes everything on the tail and is seldom your best choice. Buy the complete tail. Pictured are two peacock quills with different length and density of herl. The top quill was selected from just below the eye of the tail. The bottom herl was selected from the lower portion of the same tail feather.

3. You are looking at one side of two peacock quills (herls). It is difficult to see, but herl only covers half of the quill. Secure the quill so the bare edge is on the far side of the quill. Wrap clockwise, forcing bare portion or edge of the quill to lead the way forward.

4. *Can you see that the herl-less edge leads the herled edge forward?*

5. *Larger flies may require an underbody to achieve the desired taper or proportion. I like dubbing because it is soft, easy to taper, and peacock easily conforms to its shape.*

6. *Peacock should look like the top fly—thick and dense. The bottom fly was wrapped with the bare portion of the quill following, or over-wrapping the dense side of the herl. Notice the direction the herl faces. The top fly angles toward the right, the bottom toward the left.*

Soft Hackle Peacock

1. *Select a well marked, even-tipped partridge feather. There are two methods for installing this style of feather. Pull back fibers at tip which is tied onto the fly to create the "hackle" or legs.*

2. *Another method is to tie feather in by its base, but this shortens working length, doesn't allow full utilization of usable fibers, and the thicker stem toward base creates more bulk, especially in longer feathers. Let's use the method pictured in Photo 1.*

3. *A peacock body reinforced with fine wire has been prepared. Installing the partridge hackle completes this simple, yet effective and enduring fly.*

4. *Tie the partridge in tip-first at the very front of the thorax and back-most section of the thread head area. Remember to reserve five percent of the hook shank solely for the thread head.*

5. *Wind the partridge until the usable fibers are used up—two to four turns. Begin at very front of abdomen, wrap partridge side-by-side tightly against itself or each wrap. Tie off, taking a turn or two of thread through hackle to secure. Don't disturb the fibers as you do this.*

6. *The center stem or butt end has been cut, leaving the fibers pictured.*

7. *Stroke all fibers back away from hook eye with fingers and wrap thread slightly back over partridge to angle fibers backwards. Note the one fiber loop. I cleared this loop before completing the thread head.*

8. *Two turns of thread tied back all fibers, and they mark the position or area of the thread head. Place as few wraps as possible—just enough to cover area, finishing a small head.*

9. *Easy enough! Try tying the Olive Soft Hackle pictured on page 42 with a floss abdomen and dubbed thorax, or one of the other variations pictured. Then, continue on to the Mirage™ chapter!*

Mirage™ *(Randall Kaufmann)*

Hook: *3761BL, sizes 10-18, weight to suit*
Head: *Bead, optional, color and weight to suit*
Thread: *Black*
Abdomen: *Ultra wire, color and size to suit, epoxied over top*
Thorax: *Peacock*
Legs: *Natural gray CDC, tied fully*

Mirage™: color, size, and weight possibilities.

Only dense CDC is of value. Pictured is a good feather (left) and a poor feather (right). Learn to tell the difference.

Chapter 10

Mirage™

Angling success depends on many factors. Above all else, if fish do not see your offering, they cannot grab it. The longer it is "in the zone," the better. The fly line, presentation, water speed and depth, and the sink rate of your fly all combine to determine where the fly "fishes." Besides these factors, sink rate is controlled by adding weight to the fly, which is accomplished by wrapping wire around the hook and/or adding the weight of eyes or beads. I tie the Mirage™ unweighted, without a bead or wire weight underneath the copper wire abdomen, when I wish to fish slow, shallow water or represent emergers. If I wish to fish deeper or in quicker currents, I may add a metal bead and/or wire. For the quickest penetration and deepest sink rate, incorporate a tungsten bead and wire underneath the abdomen (essentially a double wire abdomen). The many possible weighting options allow tyers to customize the sink rate of all flies, an advantage non-tyers do not enjoy.

I like to pack my box with realistic, animated representations of various food sources and their *stages*, being certain appropriate imitations fish top to bottom. It is not practical to cover every possibility, so I have a few multi-imitators—impressionistic flies that cover many possibilities.

The Mirage™ is one such pattern, offering an illusion of midge, mayfly, caddis, stones, and craneflies, plus assorted drowned terrestrials—almost any and everything. It can be fished anywhere in the water column, lake or stream, with excellent results. Alter the size, color, and sink rate to suit your needs. The tungsten bead allows the fly to sink deeply without the aid of split shot. Not only is it easier to cast, but drifts are more natural and precise.

The Mirage™ offers practice tying a wire abdomen, working with epoxy, and installing CDC legs. The trick is to use dense CDC, tie it in by the tip, stroke the fibers backward after each turn around the hook, and use the entire feather into the fluffy butt.

Note that when patterns or instructions call for partridge or CDC "hackle," this refers to the act of winding the feather like a hackle. This is the same technique that was used to tie the legs, or "hackle" on the Soft Hackle. Sometimes the legs are split along the sides of the abdomen/thorax or pulled along the underside of the fly by manipulating the fibers with your fingers and thread.

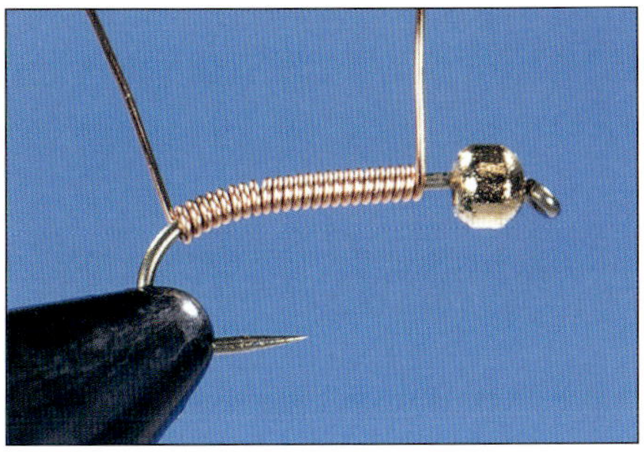

Mirage™
1. Slide on bead and wrap wire abdomen. This is size medium wire on a size 10 hook. Twist the ends off wire.

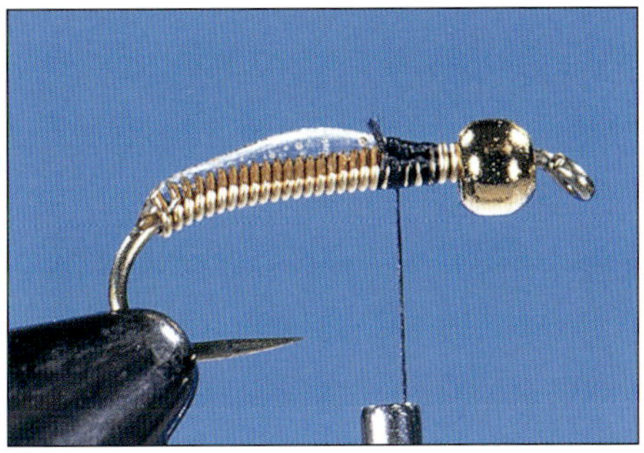

2. A drop of five-minute epoxy has been applied to secure wire; it also adds translucence and a touch of realism. Epoxy requires drying time, so make up several abdomens at a time. Bead, wire, and epoxy are in place. Then attach thread and complete.

3. Notice position and area of thread. This area becomes the thorax. Secure a peacock herl, then wrap it back to front, forming a juicy thorax.

4. It might look like the head area is crowded, but many turns of thread can sometimes be hidden under or behind the bead. Pushing thread and materials against the bead sometimes help keep it snugged tightly against the eye.

5. Select a prime CDC feather that has a relatively narrow stem and one that is fluffy at the base. The entire feather is incorporated into the fly.

6. Stroke the fibers away from the tip and tie the CDC in by its tip. Using hackle pliers, grasp the CDC and begin to wrap it around the thorax area, forming the "hackle" or legs.

7. This is also referred to as "hackle legs" (or just hackle) tied in-the-round, or 360 degrees. Stroke the fibers backwards after each wrap. This helps the fibers lay flat over the fly.

8. Wrap the entire feather and tie off. It may look a bit messy, but once the thick center stem is trimmed, the fibers are stroked back, and the thread head is complete, it should look good.

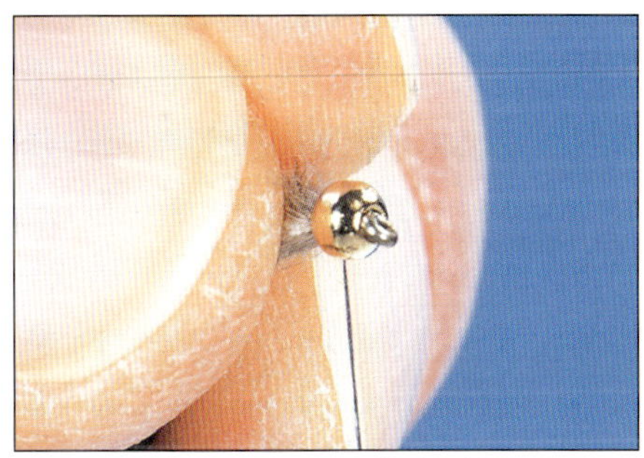

9. Remember to stroke all the fibers, using your thumb and first two fingers, which form a triangle in front of the hook eye. As they are moved back over the fly, they should capture all fibers.

10. The CDC has filled in any vacant area behind the bead, and all fibers have been tied back as the thread head is finished. Cinch two half hitches tightly, trim the thread, and lacquer.

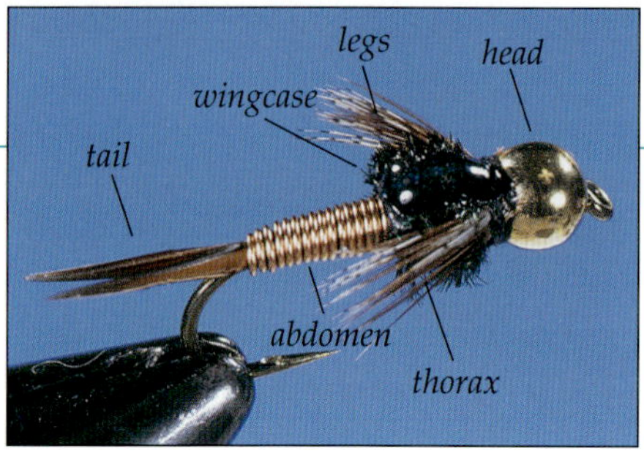

Copper John *(John Barr)*

Hook: 3761BL or 5262, sizes 10-18, weight to suit
Head: Gold bead
Thread: Black
Tail: Brown turkey biot, V style
Abdomen: Copper Ultra Wire
Wingcase: Pearl Flashabou over turkey, Thin Skin, or other base, topped with epoxy
Thorax: Peacock
Legs: Brown partridge, tied divided style

The Copper John can be tied in many sizes and colors.

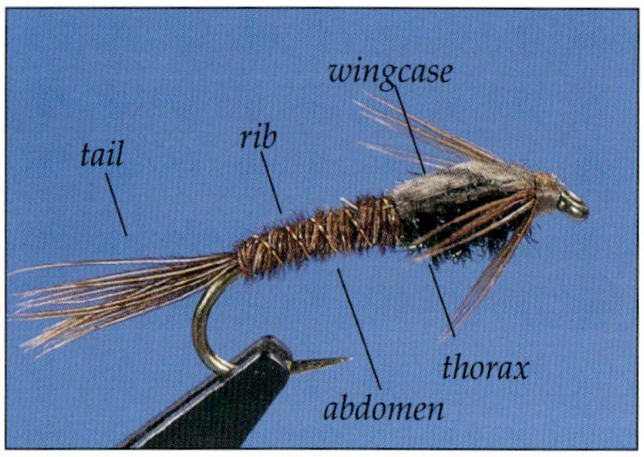

UFM

Pheasant Tail

Hook: 3761BL, sizes 10-18, wt. to suit
Thread: Brown
Tail: Brown ringneck pheasant tail
Rib: Fine copper wire
Abdomen: Brown ringneck pheasant tail
Wingcase: Brown ringneck pheasant tail
Thorax: Peacock
Legs: Brown ringneck pheasant tail

Note: The legs and the wingcase are constructed with the same bunch of ringneck fibers. After the wingcase is pulled over the top of the thorax, leftover tips are divided and secured along each side of the hook. Careful planning is needed to end up with the proper length legs.

Chapter 11

Copper John and Pheasant Tail

The Copper John was designed in the early 1990s by expert Colorado angler and innovative tyer John Barr. John needed a fly that had a rapid sink rate and attracted fish under varied water and light conditions. In this short time, it has become one of the most popular nymphs in America.

Its popularity is due, in part, to its quick sink rate which is the result of its slim profile, bead head, and use of wire. This allows it to be fished effectively without split shot, which makes for a cleaner drift with fewer tangles. In waters where fish are hard pressed, split shot often puts them on alert. The shiny copper wire, bead, and epoxy Flashabou wingcase provide the attractor qualities.

Many anglers fish this pattern like a streamer. Have you ever looked into water and noticed "glints" of light as tiny fish scatter about? The Copper John creates such reflection. This may be a trigger that fish lock onto. Try stripping a size 12 off the banks like a streamer, or swim and twitch it as it swings in the current. New techniques include the V biot tail, Flashabou and epoxy wingcase, and legs tied divided style.

The Pheasant Tail is one of the most popular and effective nymphs. Originated by Frank Sawyer in the 1930s, it has been hybridized several times. This version has all the components of the quintessential nymph—tail, rib, abdomen, wingcase, thorax, and legs. You should be able to tie it after the Copper John is mastered. The "parts" are the same but some materials are different. Many classic nymphs are tied in this fashion.

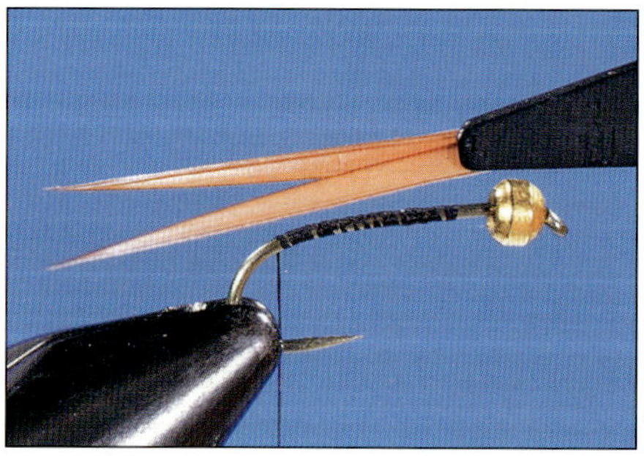

Copper John

1. Slide bead onto hook and cover shank with thread. Two turkey biot fibers are prepared in a V. Even up the tips. These become the tail, which should be as long as the hook gape is wide. Refer to page 34, photo 7, for a refresher.

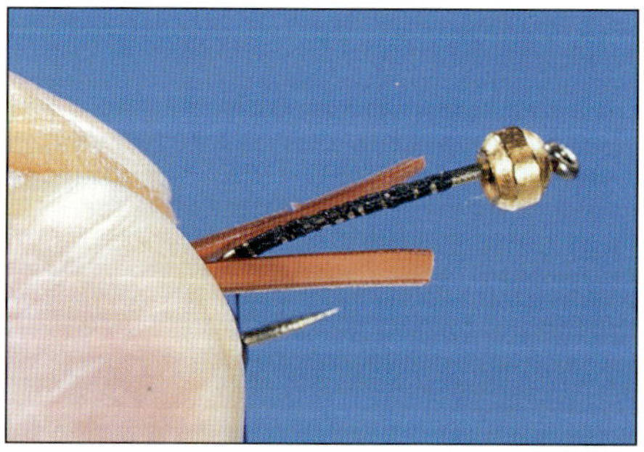

2. Then, change hands and, from back of hook, slide fibers along outside of hook. Pinch them in place and secure with thread at farthest point back on hook that you wish to construct fly, or tie on one at a time.

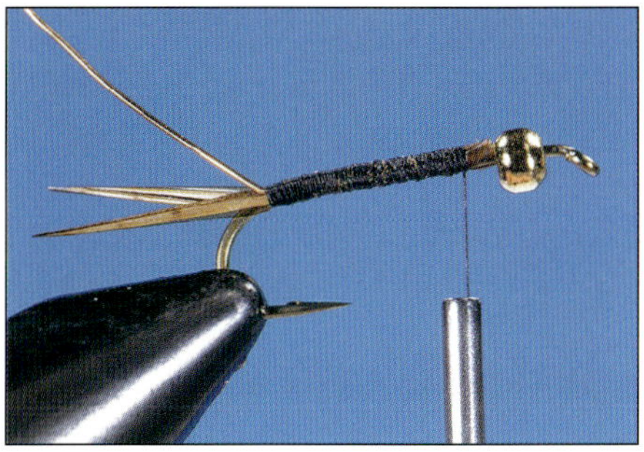

3. Tie biot down along top of hook to slightly behind bead. Beginning here, tie a five-inch section of copper wire onto top of hook. Secure back to tail area. Secure tightly at back-most position. Reposition thread at point pictured.

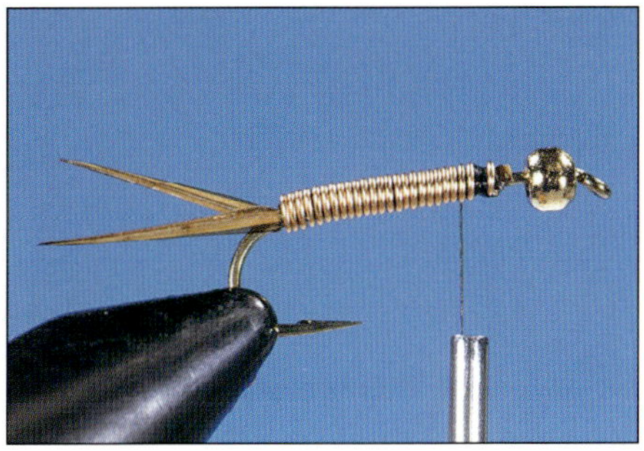

4. Wrap wire forward, each turn tightly wrapped against the last. Bring to just short of bead and tie off with several turns of thread back over the last two turns of wire. Trim or break off wire. Position thread as pictured.

5. Tie in three to five three-inch pieces of pearl Flashabou at their center point.

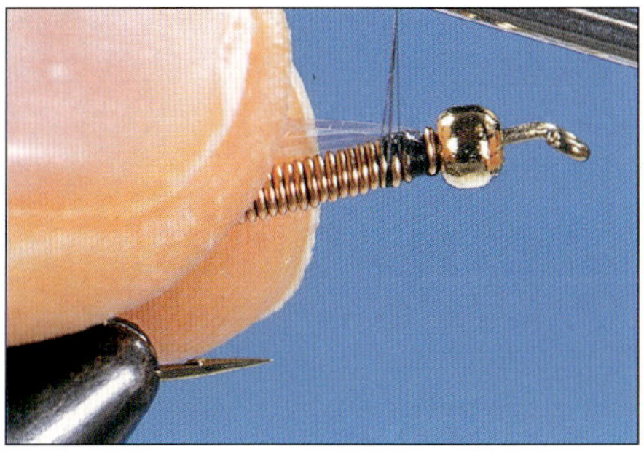

6. Double Flashabou back; tie down at back of thorax area. This reverse wrap is added security. Other materials are explained later. Trim section from white-tip turkey secondary feather; tie in. This eventually forms a double wingcase. See next photo.

7. Tie in peacock herl and construct thorax. Pull turkey forward over top of the peacock thorax with your right hand (photo 9), wrapping thread (to secure it) with left hand. Once a couple turns are in place, switch hands and tighten, wrapping with your right hand.

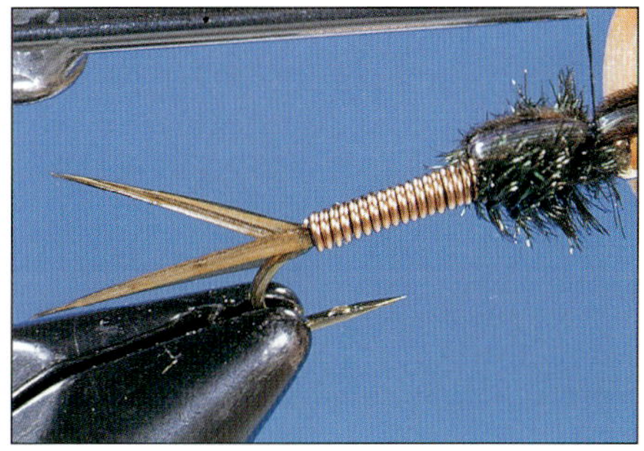

8. Pull the Flashabou over the top of the turkey in the same manner and secure. By pinching the turkey and Flashabou around the bead with your right hand, it is easy to secure it exactly over the top of the thorax and behind the bead.

9. Select an even-tipped brown partridge feather as pictured. Trim off the fluff at its base and the center stem at its tip. Holding the base of the prepared feather, stroke the fibers into a single bunch.

10. Place fibers along the opposite side of the hook with right hand. Fibers should extend about to hook point. After correct distance is measured, pinch with left hand and secure in place.

11. Once fibers are semi-secured, back off fingers so securing turns of thread can be seen. Place a few more tight turns for extra measure. Butt ends seen here must be trimmed closely away from hook eye and not protrude from finished thread head.

12. Repeat process for the near side of the hook. Measure distance and hold in position with right hand. Change hands and secure. Trim ends as shown in next photo. Finish head.

13. Place a drop of five-minute epoxy on top of the wingcase after epoxy has begun to set up, otherwise it may not stay in position. Epoxy is meant to imitate a pre-hatching distended wingcase.

14. Finished Copper John. Now, can you tie the Pheasant Tail by looking at the photo and pattern on page 46?

Plymouth Rock rooster (grizzly color), showing neck and saddle hackle. Saddle is preferred for most styles of flies if the proper size can be obtained. It offers the smallest diameter stems, best fiber density, longest length, and both stiff and soft feathers.

Two to four turns of standard hackle is about right for small to medium-sized subsurface flies. It should extend to about the hook point.

Hackle on dry flies generally extends 1½ times the hook gape.

Palmered hackle, pictured on a Peacock Simulator, generally extends 1½ times the hook gape of both surface and subsurface flies and is usually spiraled through the abdomen or thorax.

Parachute hackle, pictured on the Parachute Adams, is usually wound horizontally around a single upright wing.

Chapter 12

Hackle

For our purposes, the term *hackle* denotes domestic rooster or hen chicken feathers, either from the neck (referred to as neck hackle or simply hackle) or back (referred to as saddle hackle). Both are of value; see photo at left. The only chicken hackle to consider for small- to medium-sized flies is that grown specifically for fly production. There are several producers. Hackle is purchased on the skin. Strung saddle hackle (feathers are woven onto a strand of thread) is useful for larger flies and is usually from domestic Chinese birds.

There are four hackle choices: rooster neck and saddle, and hen neck and saddle. Rooster hackle is usually used. *Rooster necks* offer the greatest selection of feather size (size is synonymous with fiber width and denotes the hook size for which it is suitable). Necks usually have size 2, or larger, through hook size 18, or smaller. Rooster neck hackles are usually one to four inches in length. *Rooster saddles* (genetic) are mostly hook sizes 8 to 14 and up to 10 inches long, but skins usually only have two or three sizes, such as 8-10, 10-14, etc. *Strung saddle hackle* is size 2/0 to 8 and used for larger streamer, steelhead, and saltwater patterns. It is mostly five to six inches in length.

Hen hackles (both neck and saddle) are comprised of webby fibers (soft, clinging barbules) that retain water. They are short and wide, and are mostly used for wings and legs on subsurface flies (Matuka, page 53) and occasionally for dry fly wings (Adams, page 72).

There are five factors to consider about hackle. *Length:* Is the feather long enough for its intended purpose? *Size:* Is the width suitable for the intended hook size? *Fiber stiffness:* Do you require soft or stiff fibers? Notice that fibers become wider and softer toward the base, narrower and stiffer toward the top. Fibers must be stiff for dry flies and usually softer for subsurface hackles. I often like a combination of softer and stiffer hackles on subsurface flies and select and prepare the feathers accordingly. *Stem diameter:* The smaller the better; avoid stiff, brittle, flat stems. *Fiber density:* This refers to the number of fibers (barbules) along a given length of stem. More is better.

We discuss three styles of hackling—*Standard, Palmered,* and *Parachute.* All can occur on any style fly. *Standard hackle* is densely packed into a small space, usually at the front of a fly. Two to four turns are angled backward on subsurface flies. Three to 12 turns are angled forward on dry flies. *Palmered hackle* is best angled backward and is usually evenly spaced through the abdomen or thorax (page 60). *Parachute hackle* is usually wound horizontally around a single upright wing, with each succeeding wrap of hackle taken *under* the previous wrap. For parachute hackle techniques, see the Parachute Adams (page 74).

1. To judge hackle width, bend around a hook or hackle gauge, selecting proper size. Soft fibers (left) cling together and are best for subsurface flies. Web-free, stiff hackle (right) is for dry flies and can also be used on subsurface flies.

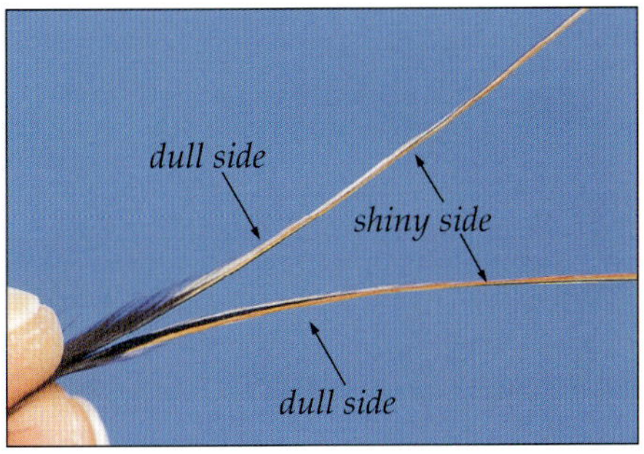

2. Hackles have a natural curvature and a shiny and dull side. The shiny side faces up on a neck or saddle skin. Here, shiny sides face together. Hold a feather on edge and visualize it tied around a hook. Which direction would the hackle slant? See photo #3.

3. Exaggerated example of angled hackle. Pictured is soft, webby grizzly color hackle. Hackle on subsurface flies should angle naturally back. Surface flies should be tied with stiff hackle angled forward (right), which provides extra support.

4. To prepare a hackle, determine where stem thickens, where fibers widen, and how much "web" you want in the fly. The web is conveniently black in all feathers on this page (except 3). Trim hackle as pictured. Full web is okay for subsurface flies, half web or less for dry flies.

5. Strip the fibers from the base of the stem. Tie hackle on edge (photo 6), slightly angling it off the far side of the hook in the direction it will be wrapped around hook. Do not tie down any fibers in the process. Tie in a bare stem as pictured.

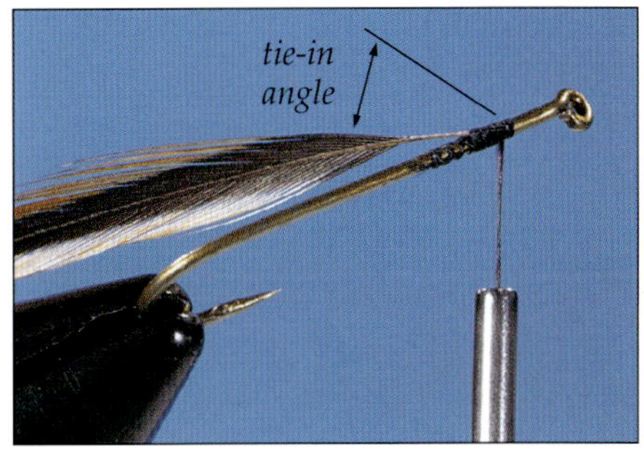

6. Whether the dull or shiny side is facing you and how the feather is wrapped determines angle of hackle curvature (forward for dry fly or backward for subsurface fly). Tie in feather on edge, angling off opposite side of hook. You are looking at the dull side in this photo.

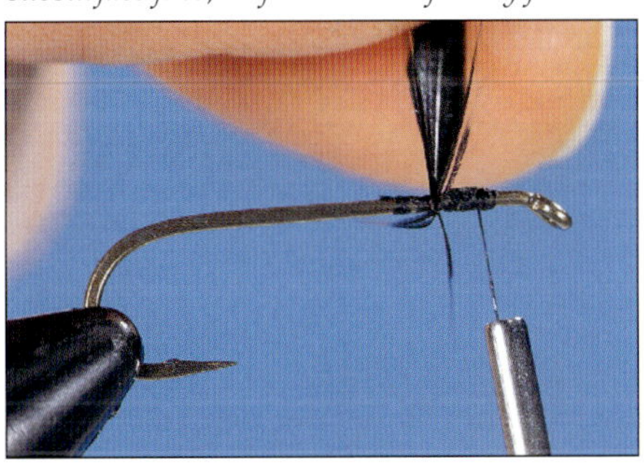

7. Begin wrapping hackle around hook so dull side faces back. Wrap back to front. Each succeeding turn is placed tightly against the last turn. Do not wrap hackle where the thread head goes.

8. Holding hackle tip in right hand, bring thread over hackle stem *as pictured, being careful not to tie down any fibers. Take a few secure turns at back of thread head area.*

*9. Grasp **all** fibers to be cut and trim closely against the hook shank. Trim all errant fibers from this area. Stroke hackle back and wrap thread slightly back over hackle to help fibers angle backwards.*

10. Finished standard hackle for subsurface fly. Dull side faces back—fibers angle back. Subsurface hackle should be softer or webby (50 percent or more web—notice black center), but this hackle is mostly web. For subsurface hackle technique, page 50.

11. Finished standard hackle for dry fly. The dull side faces forward—fibers angle forward. (It is difficult to determine fiber angle on some hackles.) Web-free, stiff hackle (25 percent or less web—notice black center) is necessary. See March Brown, page 68.

12 For palmered hackle, take two to three turns of saddle hackle at the head area, dull side to back, then spiral back to tail area. Secure with wire, wrapping forward to head area and tying off. Step-by-step directions are detailed in Woolly Bugger chapter, page 60.

Zug Bug *(Cliff Zug)*

Hook: *3761 or 5262, sizes 10-14, weighted*
Thread: *Black*
Tail: *Peacock sword, 5-8 fibers*
Rib: *Flat silver tinsel*
Abdomen: *Peacock herl (form tapered underbody)*
Legs: *Brown or furnace hackle, tied full*
Wing: *Mallard dyed woodduck, trimmed—actually a wingcase but technically, because it is not tied down at both ends, it could be termed a wing*

Note: The Zug Bug is an ever popular and regal pattern that commands attention by both anglers and fish. It is suggestive of many food forms. The Zug Bug is a refresher in peacock, provides practice installing a standard hackle and a "wing" that is actually a wingcase. Bead-Head Zug Bugs are also popular.

1. Tie in the tail. Secure the flat mylar tinsel along opposite side of hook so the silver side is facing you. When wrapped clockwise, the silver side shows. Clip the tinsel in the material clip to keep it out of your way. Wrap the peacock.

2. Secure the peacock at the front of the hook, and then wrap the tinsel rib tightly and evenly spaced throughout the abdomen. Take an extra turn at the front and tie off. Trim off the excess peacock.

3. Tie in a hackle with fibers long enough to extend to hook point when wrapped, dull side facing you. When wrapped, the dull side faces toward the back of the fly. Position thread at hackle tie-off area at back of thread head area, front of hackle area.

4. Wrap hackle back to front, placing each turn immediately in front of the last turn. Three turns is right for a size 10 hook. Tie off. Notice position of thread and small but vacant space behind eye. This is fairly stiff hackle. Softer hackle can be used.

5. Stroke all fibers backwards and wrap thread back over hackle, angling fibers back over peacock. Position thread at rear of thread head. This space ultimately becomes the thread head.

6. Select a mallard flank feather dyed woodduck color and trim out a small bunch of fibers. The tips do not have to be even as they will be trimmed to shape. This feather can make three small flies, one from each side and one from center tips, or one larger fly.

7. Hold mallard fibers in place over the top of the fly with your right hand.

8. Pinch in position with your left hand and secure in place at the rear of the thread head area. This allows for the butt ends to be trimmed short behind the hook eye so that a neat head is constructed.

9. Trim off excess at front, making one clean, close cut, and finish the thread head. Trim the wingcase so it extends one-third of the way over the top of the abdomen.

10. Finished Zug Bug. You have now learned to construct a tail, rib, abdomen, wingcase, and standard style hackle, and how to work with dubbing, peacock, marabou, wire, and hackle! You can now tie many, many flies!

Tying Streamers

Streamers often represent fish. Some are designed to imitate specific fish—chubs, dace, sculpins, bass, salmon, trout fry, etc. Like nymphs and dry flies, others are designed to be exciter, attractor, or impressionist patterns. Open any angler's fly box and odds are, there are only a few token streamers. They are the most overlooked and underfished flies of freshwater anglers. Considering the fact that most big fish dine heavily on little fish and the goal of many anglers is to fool big fish, it is only logical to fish with streamers. Saltwater anglers know this best, realizing that imitations of fish are the password; they stock their boxes accordingly.

Some streamer patterns are simple, others complex. We are tying three patterns: Matuka, Clouser Minnow, and Marabou Muddler. The Matuka and Clouser Minnow are relatively simple *styles* of streamers that can be customized. The Marabou Muddler is a bit more difficult, but it provides some basic instruction about working with, spinning, and trimming hair.

The Marabou Leech, described in chapter 8, and the Woolly Bugger, described in chapter 15, could be termed crossover patterns. Both are fished as a representation of leech, and perhaps larger nymphs, but they can also represent fish (streamers) and be termed exciter, aggravator, and attractor patterns.

Because there are so many different styles of streamers, no hard and fast rules exist for proportions. Aside from specific minnow imitations, the single most important consideration is wing length. Too long and it catches or fouls in the hook bend. Too short and its animation and silhouette are reduced.

Fly tying is not mathematics. There are few, if any, rules. Deviation from the norm, imagination, and innovation prevail. Its no wonder the sport attracts so many misfits who are at their best when playing alone and out of bounds.

If any fly patterns fit into a specific mold and also break all the molds, they are steelhead flies. Size, color, and style span much of the tying spectrum. They are basically a streamer style fly but they are referred to as steelhead flies.

We are tying basic steelhead flies that have a tail, rib, body, hackle, and wing. There are hundreds of steelhead and Atlantic salmon patterns with these parts in common. Pick a material and color combination. There isn't anything revolutionary about steelhead flies, but because they are beautiful to look at, fun to tie, and allow your imagination to run wild, steelhead flies are the favorites of many tyers. Steelhead fanatics have hundreds of flies neatly arranged in their many fly boxes and wallets that are truly beautiful and fascinating to look at.

Streamer Proportions (300)

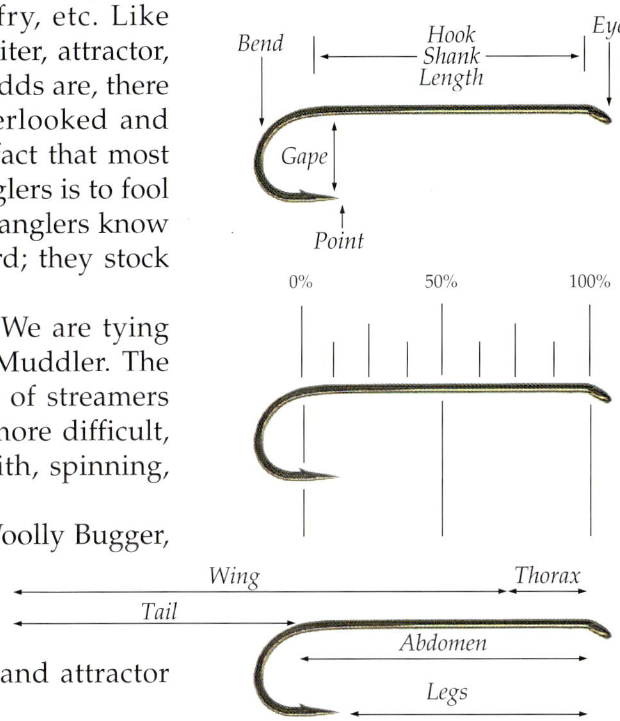

Streamer-style flies vary greatly and so do their proportions, but the following are general guidelines:
Tails can extend as far as the hook shank is long but are usually shorter.
Abdomens usually cover the entire hook shank.
Wings can extend up to twice the length of the hook shank.
The thorax usually begins at the 70-80 percent mark and continues forward to the hook eye.
Legs (collar, throat, hackle) can extend to the hook point or less.

Always smash barbs when hooks are not barbless.

Black Matuka
Hook: 5263BL, sizes 4-8, weight to suit
Thread: Black
Rib: Fine wire
Abdomen: Black dubbing
Thorax: Also called a "throat"; red dubbing
Wing: Black hen hackle, two feathers face to face tied Matuka style, and black and red Krystal Flash
Hackle: Black saddle
Eyes: Yellow/black, painted onto head, optional

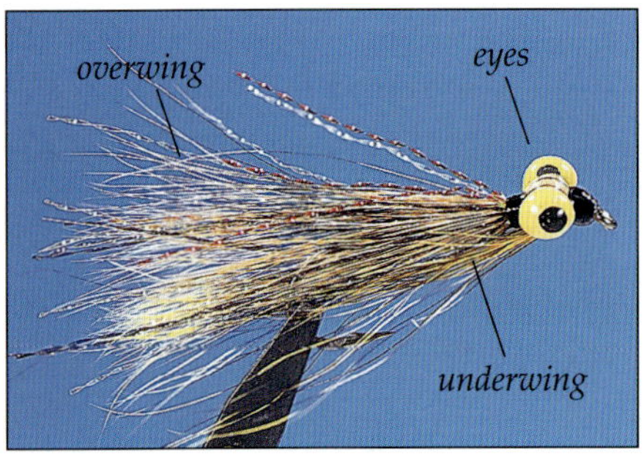

Clouser Minnow (Bob Clouser)
Hook: 5263BL, sizes 4-8
Thread: Black
Eyes: Barbell, painted to suit
Overwing: Red fox squirrel hair with gold, red, and pearl Krystal Flash
Underwing: Red fox squirrel dyed yellow

Marabou Muddler
Hook: 5263BL, sizes 2-8, weight to suit
Head: Gold conehead, optional
Thread: Black
Tail: Red hackle fibers
Abdomen: Silver holographic tinsel
Underwing: Natural brown bucktail or substitute brown calf tail
Overwing: Any color(s) of marabou, silver holographic Flashabou, and red Krystal Flash
Topping: Peacock (not shown)
Collar: Deer or caribou hair
Head: Trimmed deer hair

Chapter 13

Matuka, Clouser Minnow, and Marabou Muddler

The Matuka is a *style* of streamer whereby the wing is tied down on edge, along, or over the back of the fly. There are all colors of Matukas and they can be tied to represent any fish, such as shiners, shad, sculpin, roach, tadpoles, etc. The Black Matuka could imitate various darker fish, leech, eels, etc. It provides practice weighting a hook, creating a shaggy dubbed abdomen and thorax, and installing standard hackle and a matuka wing. This wing is a hen hackle feather, but it could be any feather. Rabbit or other fur on-the-hide is often used.

Following the Matuka is the Clouser Minnow, a fly that has fooled more species of fish in more waters than any other. Developed by smallmouth bass expert Bob Clouser, it belongs in every angler's arsenal—saltwater, bass, trout, exotics, etc. Remember, all fish love to eat fish and the larger the fish, the bigger their prey. The tying technique is unique to this pattern, but it can be adapted to represent any species of fish. It also offers practice "stacking" hair.

Next is the classic Marabou Muddler, which is a variation of the even more classic Muddler Minnow. Developed in the 1950s by Don Gapen, the Muddler Minnow revolutionized the tying style and shape of streamer flies. Today, its influence is found in many popular patterns.

The Marabou Muddler was popularized by Dan Bailey and Joe Brooks in the 1960s and continues to reside in fly boxes throughout the world. It is a "big fish" fly that can be tied in many colors, combinations of colors, and various sizes and weights.

New techniques include creating a tinsel body and bucktail underwing. The stiffer bucktail underwing helps keep the softer marabou overwing in position and provides a little mottling effect. Calf tail is often substituted for bucktail on this and other streamers. The marabou wing is like the previously tied Marabou Leech. The deer hair collar and trimmed deer hair head are more difficult. Select moderately soft deer hair that flares when it is tied down, but still has some dark color in the tips. Surround the marabou wing with a bunch of "stacked" deer and tie it down at the immediate front of the marabou wing. This forms the collar. Next, select another bunch of deer hair, trim the tips, and "flare" it in front of the collar. You may need to do this two or three times to fill in all the space between the conehead and wing. Work with small bunches of hair to start. Wrap the thread tightly through the hair, tie off the thread, and trim the hair head to shape.

Black Matuka
1. First, cover the hook with thread, then add desired amount of steel wire. Then tie in the fine wire that becomes the rib, and finally, begin dubbing the abdomen. This is scraggly, somewhat robust dubbing.

2. Work a taper into the abdomen. Either work back to front with an increasing amount of dubbing on the thread, or double back, then forward, or begin at center point, wrap forward, back, forward. The red "throat" or "gills" adds contrast, and trout like red.

3. These two hen hackles become the "Matuka" wing. To even the tips, face the dull (concave) sides together, essentially forming a single-looking flat feather free of curvature.

4. Trim the fuzzy butt off at the length you want the wing to be, usually twice the hook length.

5. Tie the feathers onto the hook at this point, being certain the feathers remain together on edge over the top of the hook. The wire rib ties the feathers down at the back of the abdomen.

6. Spread the fibers at the tail tie-down area where the first turn of wire rib goes. This makes it easy to secure the feathers without creating errant fibers and twisting or rolling the wing out of alignment. Pull wing tight and secure.

7. The first turn of wire rib is in place. Continue "ribbing" through the abdomen, being careful to keep the wing square or straight across the top. Do not tie down any fibers.

8. The rib and wing tie down is complete. The thread position pictured is where the hackle should be tied in place. Hackle and the thread head fill in the space left. Note that space for hackle and a vacant area behind the hook eye remains to build the head.

9. Add red, pearl, and black Krystal Flash along sides. Tie hackle at front of "gills." Wrap hackle back to front, each succeeding turn placed immediately in front of the last. Leave no space between wraps. Finished hackle should angle back, dull side to back.

10. Hackle has been tied off, but the tip has not yet been trimmed off in this front view. Trim tip and finish fly.

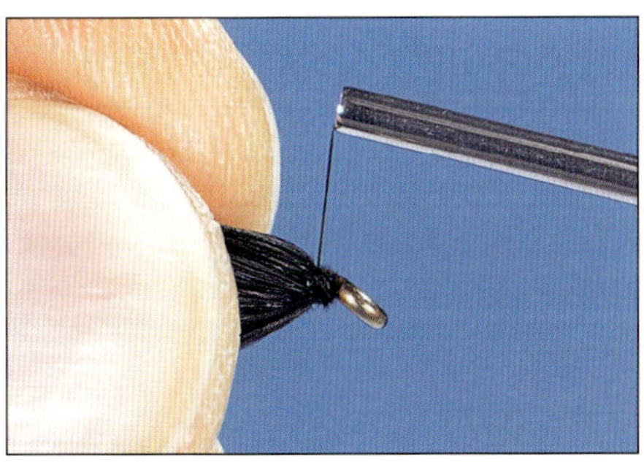

11. Stroke all fibers away from the head area and finish a neat tapered head.

12. Finished Black Matuka. Notice proportions, angle of hackle, tapered abdomen, and silhouette. If you were a fish and this came into your territory, wouldn't you be tempted to eat or at least attack it? Many have!

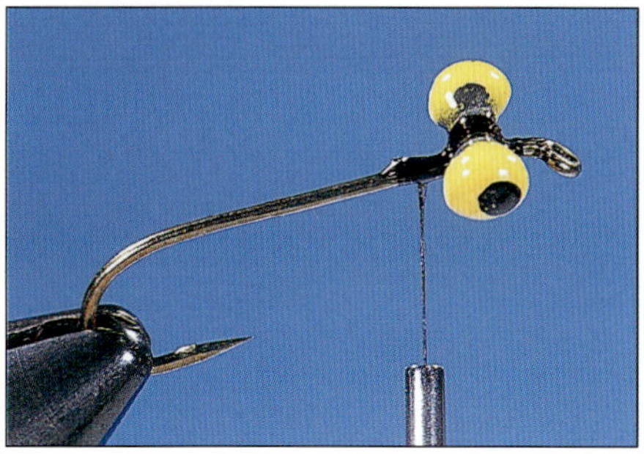

Clouser Minnow
1. Secure barbell eyes, size to suit onto hook at 85-90 percent mark. These are extra-heavy eyes (paint or purchase pre-painted). X-wrap between, around, and underneath the eyes. Zap-a-Gap helps secure eyes in place.

2. Trim off a bunch of squirrel hair, which will be "stacked." A stacker is a take-apart tool indispensable for evening the tips of hair.

3. Before stacking, hold hair by tips and pull out all shorter hairs, reducing bunch by 20-30 percent, which facilitates "stacking."

4. Place the natural tips (end to be stacked) into stacker. Place index and third finger over top of stacker and tap bottom on a hard surface. Straight moose body hair stacks with one or two taps, while crinkly calf may require over 30.

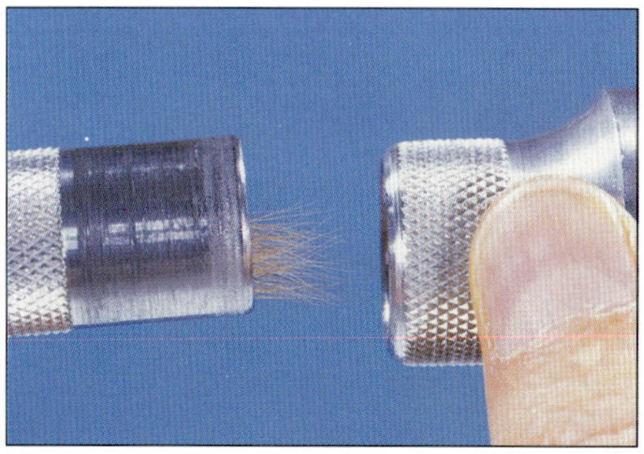

5. Angle stacker up slightly and slowly remove end, revealing stacked hair. Remove tips with right hand, then hold butts in left hand. Pull away any "short" fibers with right hand.

6. Hold butt ends with right hand behind hook eye. Hair extends over and between eyes. Measure length of wing. Pinch hair tightly around top of hook eye. Tentatively secure with three to four turns of thread with left hand.

7. This helps begin tie in. It is difficult to hold and tie these "hard" somewhat large-diameter fibers in the usual left hand to secure because of tight space between eyes and hook eye, so we begin as per photo 6, then change hands to finish tie-down.

8. Once secure in front of eyes, secure behind eyes as pictured, tying down the top wing. Trim butts well behind or away from hook eye. Tie in Krystal Flash along underside, bottom, top, or sides of wing.

9. Repeat steps two through five and secure another bunch of hair along the underside of hook, exactly as you did over the top. If you have a rotating vise, this is easier. Start tie in, holding hair with right hand.

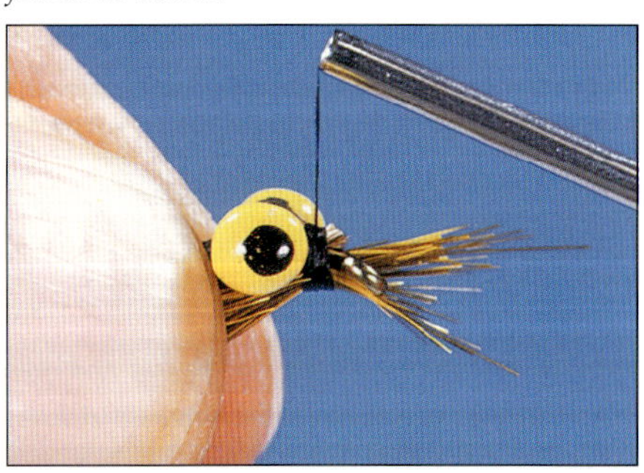

10. Switch to left hand, holding hair, and wrap thread with right hand, finishing a secure tie down. Bottom wing should be same length as top. Tie in close to eyes so there is room to trim fibers behind hook eye.

11. Notice tie in and trim of bottom wing. If wing is not kept back from the hook eye, it is not possible to trim fibers as pictured, leaving space to finish the thread head.

12. Finish off the head and lacquer. Epoxy makes a slick, smooth finish that should outlast the fly. Note Krystal Flash in wing. The Clouser can imitate any baitfish or be tied in any "attractor" color.

Marabou Muddler

1. Slide conehead onto hook. Tie in the red hackle fiber tail, then tie in a five-inch section of medium to large (depends on hook size) flat tinsel. Wrap tinsel back to tail and back to cone and tie down.

2. Select several fibers of natural bucktail or tan calf. Stack the tips evenly in a stacker and tie in behind cone. A loop of thread can be taken under and around wing to help bunch it together and stand it up. Trim butt ends as pictured.

3. Select marabou for wing. This is one black and one white plume with the center stem trimmed out. Bunch the fibers together so the tips are even and secure over bucktail underwing.

4. Marabou wing is in place. Lash down securely directly on top of the hook. Length extends about to end of tail. Tie in Flashabou, Krystal Flash, etc. for wing highlights (see photo 5) and peacock topping, (not shown) if desired.

5. Select a clump of medium soft deer hair that flares when lashed down to hook. If it doesn't flare, you can't get the trimmed effect or pack the fibers densely enough. Stack tips evenly. Hold in right hand to measure distance.

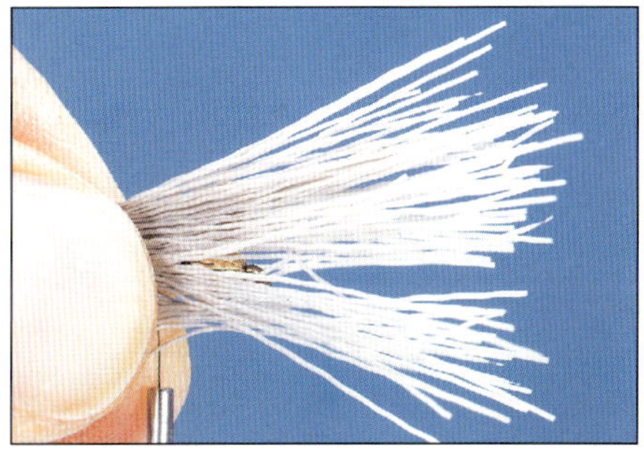

6. Change grasp to left hand, pinching fibers immediately in front of wing. Tie in at immediate front of wing. Cinch the thread slowly as this will probably test its breaking strength. Use heavier thread if needed.

7. The deer hair collar has been tied in place, encircling the hook 360 degrees. The butt ends are a bit long and difficult to control. Trim them by 50 percent. These fibers later become part of the "spun" and trimmed hair head.

8. There should be empty space between this tie-in and the cone. Stroke all the butt ends back away from the cone, tying them back, making room for the remaining hair that forms the head.

9. Trim a section of soft deer or caribou. Prepare it as shown—tips have been trimmed off—and hold it in your left hand. Note that caribou is perfect for this type of "hair spinning" but not for a collar because it is too soft and colorless.

10. Place hair as pictured and tie down as shown. Easy does it! Allow hair to roll and spread out, which is good. Sometimes one clump fills in 360 degrees around the hook; sometimes it takes two or more clumps.

11. Wrap thread tightly through the hair so it disappears to just behind cone. Finish off knots, trim thread. Be certain thread wraps penetrate the hair to the hook; otherwise, it shows or gets cut in the trim process.

12. Razor blade hair to shape. Continue taper of cone. Do not trim collar. All "spun" hair flies—Irresistible, Muddler, bass bugs, etc.—are tied like this. It takes practice, but you have an idea of how to do it!

Hot Butt® Steelhead *(Randall Kaufmann, 2000)*

Hook: 7999 or 7989, sizes 2-8
Thread: Black
Wing: Mirror Flash, color of choice
Butt (Tail): Glo-Bug yarn, cerise, Alaskan roe, or color of choice
Rib: Flat or oval silver tinsel
Abdomen: Ice Dub or STS Trilobal black, blue, or color of choice
Hackle: Color of choice
Note: A new-generation fluorescent, low water, "shimmer" pattern. Black, purple, blue, and orange are good base colors.

Green Butt Skunk *(Dan Callaghan, 1957)*

Hook: 7999, sizes 2/0-8
Thread: Black
Wing: White calf tail
Tail: Red hackle fibers
Rib: Flat silver tinsel
Butt: Fluorescent green chenille
Abdomen: Fluorescent green chenille, black chenille
Hackle: Black
Note: Perhaps the most popular steelhead fly in the Northwest.

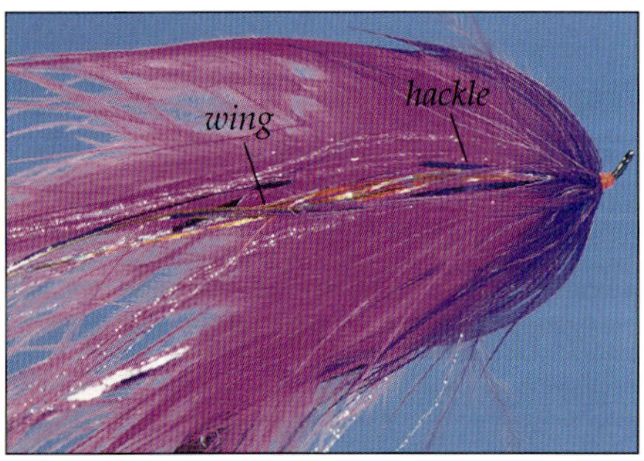

Alaskabou Showgirl *(George Cook, 1985)*

Hook: 7999, sizes 3/0-2
Thread: Fluorescent fire orange
Wing: Marabou "hackled" and/or tied 360 degrees with several Krystal Flash, Holographic Fly Flash, and Flashabou highlights; I like pearl, holographic, red, fuchsia, gold, fluorescents
Hackle: Purple; wide and soft
Note: Use your imagination—concoct your own personal favorite combination of colors.

Chapter 14

Hot Butt® Steelhead, Green Butt Skunk, and Alaskabou

Most steelhead flies are relatively easy to tie. If you can tie the basic style consisting of a wing, tail, body, rib, and hackle, you can tie most steelhead and many Atlantic salmon patterns. Tying technique remains the same; only the colors and materials change.

New technique includes the installation of a reversed wing and hackle. Reversing the wing and hackle insures that they do not pull out. The reverse wing and hackle create a "no-brainer" perfect head. All other tying techniques required to tie these flies have been discussed previously.

The design of steelhead flies has evolved over the last 40 years to less is better, meaning dressings have become more sparse. The amount of materials used on a steelhead fly is subjective and can depend on water conditions and personal aesthetics. Generally, the bigger, darker, and colder the water, the bigger and more heavily-dressed the fly should be. Winter anglers often use larger flies than summer anglers, who often prefer a sparse "low water" dressing.

We are tying three flies, a Hot Butt® Steelhead, a Green Butt Skunk, and an Alaskabou. The Hot Butt® is considered a "low water" contemporary dressing and presents a slim, shimmering silhouette of many colors. It is animated, highly visible, and attracts the attention and curiosity of fish from a great distance.

The Green Butt was designed by famous Oregon angler/photographer Dan Callaghan for the North Umpqua River during the 1960s. Dan is a regular on the river with camera and fly rod and hangs out in Cabin 1 at Steamboat Inn. The Green Butt described here is considered a standard dressing for today, not too bulky, but not low-water style.

The Alaskabou, designed by George Cook, has it all—simplicity, angler appeal (they look beautiful lined up in a fly box), popularity, personality, and versatility. The marabou slims down when wet and comes to life in easy currents and/or with a twitch of the rod tip. The Alaskabou excites all gamefish, including steelhead, char, Dollies, trout, and all species of salmon. The cerise/purple version (Showgirl) is favored by silvers; try black/blue or pink for kings; white/green for chums, sockeye, and kings; white/red for pike; white, black, or olive for trout. These are great fun to tie and fish. The Alaskabou demonstrates installing marabou as "hackle" and installing standard hackle with a fluffy base.

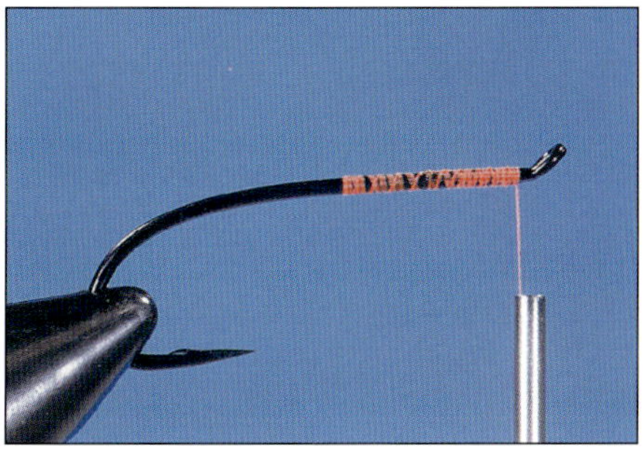

Hot Butt® Steelhead

1. Practice finishing a reverse wing head. Steelhead wings are tied reverse whenever possible. Position red thread (for visibility) exactly as pictured.

2. Select 12 Mirror Flash fibers (I doubled over six), secure so long ends extend forward. Note exact position of thread. From here forward must be reserved for head. Pretend the fly has been completed and finishing the head is all that is left.

3. Since you reserved this space, let's complete the thread head. Pull the fibers flat and tight back over top of hook. This is the wing.

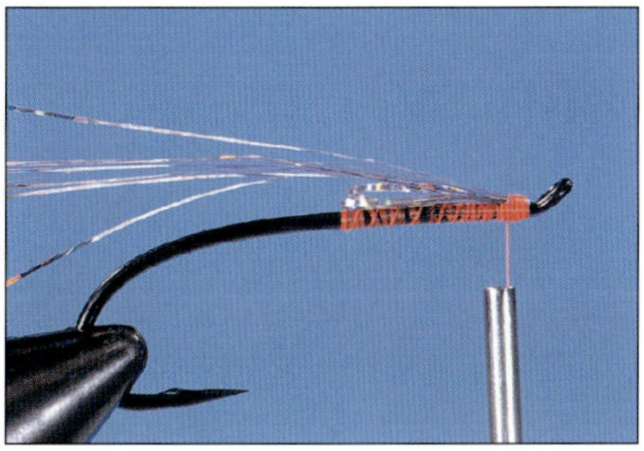

4. Take six-eight tight turns of thread, securing the wing in a backwards position and forming a perfect head. How can you miss? This thin wing material easily makes a tiny head. Hair is bulkier. Undo the head and continue tying.

5. Cut off a length of Glo-Bug Yarn and secure and trim exactly as pictured. Tail begins in front of the hook point. Remember, this is a sparse, low water tie. Note angled cut of yarn at front, and the position and number of thread wraps.

6. Tie in the tinsel rib and dub a thin amount of dubbing onto thread. Notice how sparse the dubbing is. You should see thread through it. Keep it sparse! We changed to black thread, but red is okay, too.

7. Install the dubbing, then the rib to the point pictured. Remember that the space for the thread head is off limits and that we need space for hackle.

8. Another method of tying in standard hackle is to reverse it, much like the wing. Hackle extends forward, dull side away from you. By doing this, hackle does not pull free at any time. Hackle fills in the black thread area. Half of red area is reserved for thread head.

9. Two-three turns of hackle are wrapped front to back, and hackle tip stem is tied down at the back. Wrap thread through hackle, front-back-front. Do not tie down any fibers. Trim or pull off tip of hackle.

10. Trim the butt end of the hackle stem. You can see that the reverse wrap is very secure and clean to do.

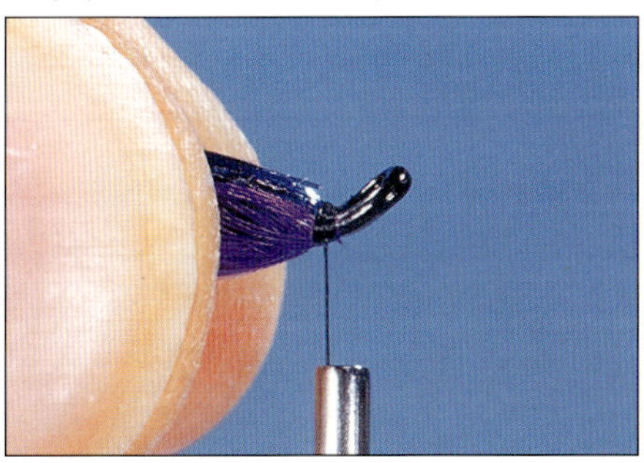

11. The last step is to pull the wing back over the fly and tie it down (just as you did earlier) and complete the thread head. If you reserved the space, it should look the same.

12. Completed Hot Butt® Steelhead. Vary the colors and fill up a fly box. They look real cool all lined up, and they look even better swimming like a waving neon banner in the water! I like black, blue, purple, orange, and fluorescents.

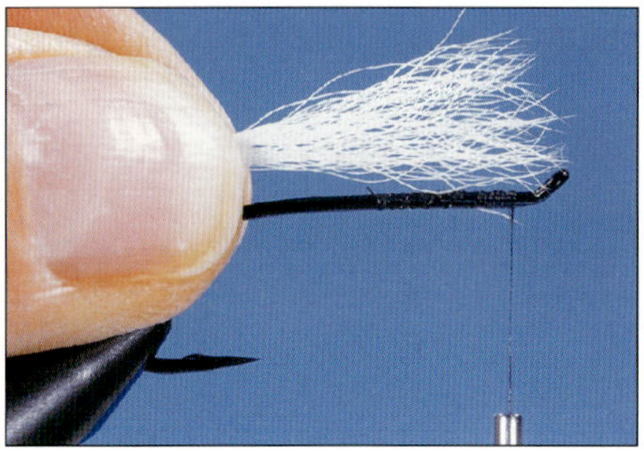

Green Butt Skunk

1. Select and stack calf tail. When tying a reverse wing that you can't trim to length it must be right when you tie it in. It should be length of hook shank. Pictured is right length. Move fingers forward and tie in place.

2. Calf wing is secured same as we did with the holographic Hot Butt® wing. Today standard style steelhead flies usually have a tail about half the hook length that begins perpendicular to hook point. Position thread as pictured at tail tie-in area.

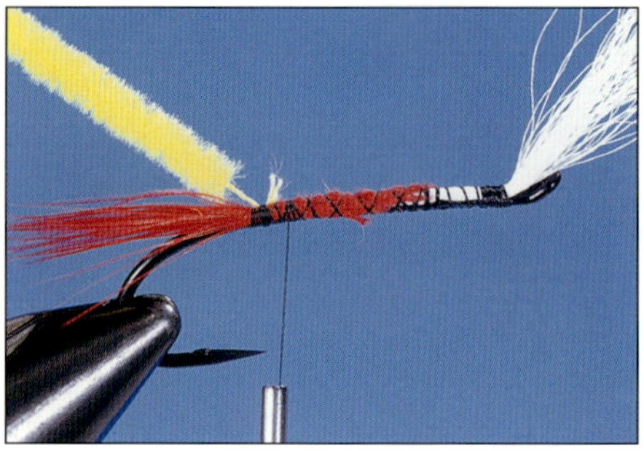

3. Tie in tail so it melds into wing tie down. Twist off ends of chenille and tie in place. Two-three turns of this chenille form the butt. Position the thread at the point pictured to tie off the chenille butt.

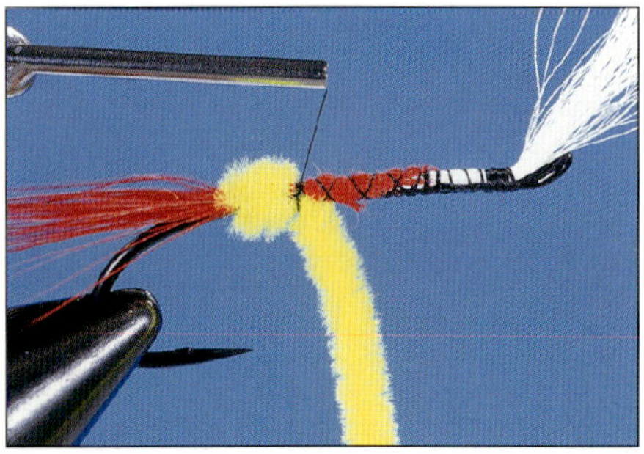

4. Wrap butt and tie off. Trim ends closely.

5. Twist off ends of black chenille and tie in. Position thread at front where body is to be tied off. Remember to plan space for body tie off, hackle, and thread head.

6. Chenille and rib are in place. Rib must be tight. Space it evenly. Take an extra turn at front of body so it is easy to secure with thread. Notice rib spacing.

7. Tie in hackle with dull side facing you. If you elect to tie in a reverse hackle, the shiny side faces you.

8. Take three-four turns of hackle and tie off, being careful not to intrude on thread head area.

9. Pull wing back tightly and flat and take a couple of turns of thread over top of reverse wing. Check it out. How does it look? Secure wing with more thread and finish a smooth, tapered head.

10. Epoxy the head, and the Green Butt Skunk is finished! Today, most steelhead flies have some reflective material in the wings, either Flashabou or Krystal Flash.

11. Green Butt Skunk tied low water style. Bodies are dubbed sparsely. This is basically a size 8 fly tied on a size 4 hook or a size 6 on a 2, etc. Lighter wire 7989 hooks are often used.

12. Many "reduced" Atlantic salmon flies are similar in construction. Pictured is a Ponoi Green (top) and a Silver Rat. Could you tie these by sight without a pattern?

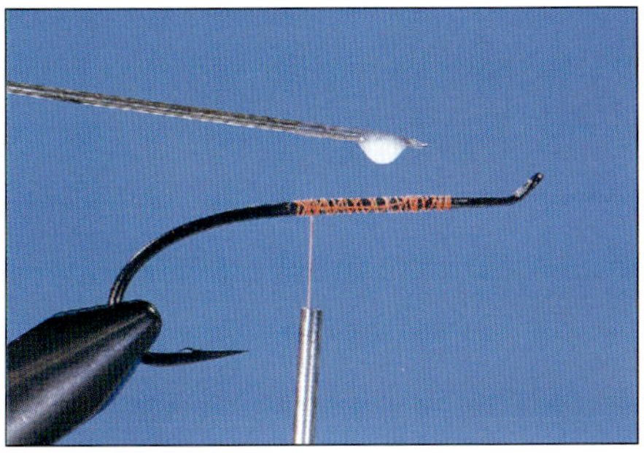

Alaskabou

1. Cover the front 70 percent of the hook with thread and lacquer. This is necessary to help keep the bulky materials from sliding on the hook.

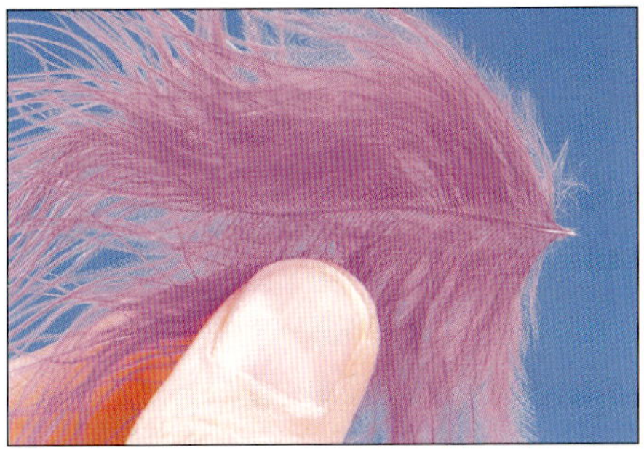

2. Trim the top portion of a marabou plume, eliminating that portion with the thickest stem. The thick stem can be brittle, it does not bend well or wrap easily, and it takes up too much space.

3. Just like hackle, there is a shiny and dull side to marabou, only it is very subtle and it is not imperative that it be tied one way or the other because it is so whimsical. This is the dull side, and it faces back when wrapped in place and fills the space to the position of the thread.

4. Wrap it just like hackle. Because fibers are so long and soft, you must constantly "clear" and stroke them backwards. Hold onto the very tip and wrap as much of the feather as you can.

5. We have wet the marabou for better control and visibility of the tying area. Tie off and trim the marabou tip.

6. Repeat the process, tying in another section of marabou. Wrap it, and tie it off.

7. Second marabou plume is installed. This pattern can also be constructed by tying in clumps of marabou, one after the other, encircling the hook, but it does not create the same effect.

8. Let's finish the wing with three clumps positioned around the hook. This beefs up the shoulders and gives it more definition. Remember, you look at these flies more than fish do!

9. Three clumps have been tied in place. Note the spacing left to tie in the reflective highlights and hackle. The reflectives can be tied in individually or as a group.

10. The group tie in is easiest. Cut seven-inch long lengths of fibers (this is a three-inch long fly on a 2/0 hook) and place them on the table. Tie half the length along one side of the hook, half on the other.

11. Install a long, wide, soft hackle, using as much of the fluff at the base as possible. Finish off and lacquer the head.

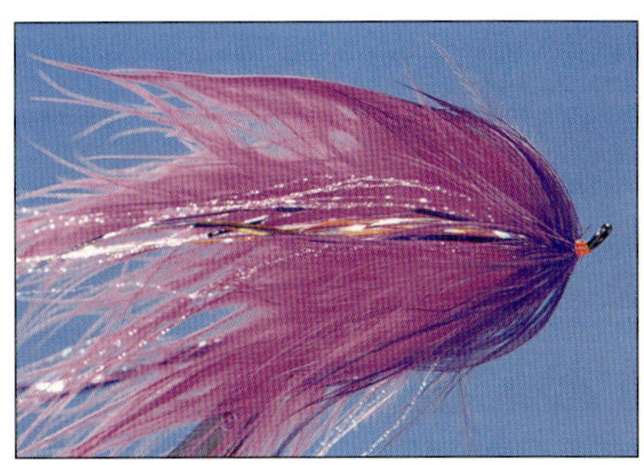

12. Finished Alaskabou. Collect several colors of marabou and reflectives and stock your fly box. This is a great streamer for all gamefish everywhere. Customize the colors.

Woolly Bugger
Hook: *5263BL, sizes 4-10, weighted*
Head: *Bead; color, weight, and style of choice*
Thread: *To match abdomen*
Tail: *Black marabou or color to match or contrast with abdomen, and a few strands of Krystal Flash, etc.*
Rib: *Gold wire*
Abdomen: *Black chenille; fur mixes also good*
Hackle: *Black saddle hackle, or color to match abdomen, palmered*

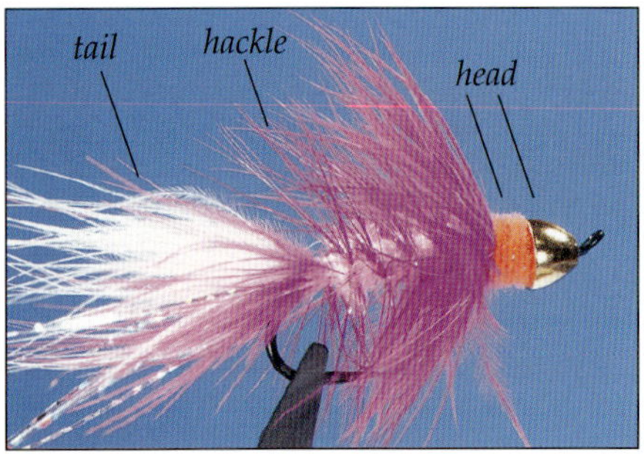

Misfit™ *(Randall Kaufmann, 1996)*
Hook: *7999, size 4, weight to suit*
Head: *Gold cone*
Thread: *Fluorescent fire orange*
Tail: *White and cerise marabou and Holographic or Mirror Flash highlights*
Rib: *Silver wire*
Abdomen: *Pearl Crystal Chenille*
Hackle: *Cerise, palmered*
Head: *Fluorescent fire orange chenille*
Note: Other colors to suit-black, purple, pink, orange.

Chapter 15

Palmered Hackle Techniques

Woolly Bugger and Misfit™

The Woolly Bugger has become one of the most fished and effective flies in the history of fly fishing for trout! It may represent a multitude of foods, including leeches, minnows, salamanders, tadpoles, eels, snakes, crayfish, craneflies, dragons, stones, and pancora in Chile and Argentina. It could be classified a nymph, streamer, leech, or crustacean. No other pattern compares as a trout exciter, and fish are just as likely to attack it out of aggression or curiosity as they are for food.

It can be tied in any color with many materials. Black and olive are the standbys, but I like red, maroon, white, brown, orange, fluorescents, and combinations of colors. Add a bead, a conehead, eyes, or rubber legs for animation...use your imagination.

Note tapered hackle on the finished Bugger. This is why saddle hackle is tied by the butt end and wrapped front to back. The saddle was trimmed just above the fluffy fibers, which are the widest. Three turns were taken in place just behind the bead, and the feather used right to its tip, where fibers are more narrow, creating a perfect taper.

The Misfit™ is basically a Woolly Bugger with the addition of a conehead for weight and a chenille "head." The flashy cerise color scheme is one preferred by steelhead and Pacific salmon, especially coho (silvers). It could be termed an attractor or exciter streamer, but it most likely owes its effectiveness to its resemblence of salmon roe or eggs.

Palmered Hackle
1. Secure wire rib. Dubbing is for effect. Select proper width saddle hackle (1½ times hook gape). Trim off fluff at base, strip fibers from stem, and tie in on edge, angling off opposite side of hook, dull side facing you.

2. Begin with two to four turns of hackle in place at head area, then spiral evenly toward tail. Secure with two turns of wire and continue wire forward as rib. Weave wire through hackle. Be careful not to matt down fibers.

3. Take extra turn of wire at front and secure with thread. Leave space for the head. Trim leftover hackle tip, being careful not to trim fibers, or, if secure enough, pull off with fingers. Gather fibers and pull back away from hook eye and finish the thread head.

4. *If your hackle looks like the upper, continue. If it looks more like the lower, try it again. Take your time, prepare, and tie in hackle as pictured. Watch thread placement and keep constant tension on hackle.*

Woolly Bugger

1. *Depending on the hook size in relation to the marabou plume, one or more may be needed. Trim out the center stem and even the tips. Bugger tails do not necessarily need the center stem removed.*

2. *If using complete plume, one is usually enough. Draw wet fingers through plume to better control it. Measure marabou length against hook with plume in right hand, change hands, and secure over/along top of hook.*

3. *Bead, wire weight, and marabou are in place. Note positions: wire toward front, space behind bead to complete fly, marabou secured behind wire about length of shank, and level foundation.*

4. *Tie in fine wire, which becomes rib and secures hackle. Trim five-inch lengths of a few flashy materials (holographic Flashabou, Mirror Flash, or Krystal Flash) and tie onto near side of hook along tail.*

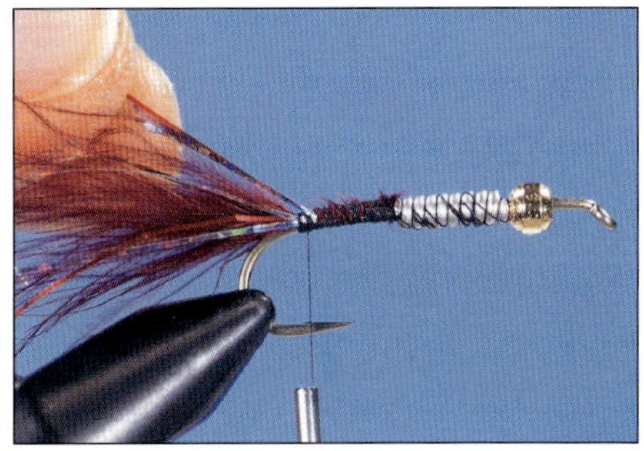

5. *Pull ends of flashy materials over to opposite side of hook and secure so they extend along opposite side of tail. This saves making two material selections, two tie ins, and two cuts. Trim to about length of tail.*

6. *Chenille is a bit bulky to tie on without first untwisting it down to the thread core. If you do not untwist it, tie it down along the entire abdomen, which increases fly diameter, but keeps a level foundation.*

7. *Tie in the thread core. Position the thread at the front of the wire weight. Wrap chenille forward to just behind bead, forming abdomen; tie off. Prepare hackle. I like a little web, especially at the feather's base, which creates a soft, wider collar at front of the fly.*

8. *Tie in the hackle, dull side facing toward you, angled away off the opposite side of the hook. The first turn is taken at the back of the soon-to-be thread head area, dull side facing toward the tail.*

9. *Take three turns in place at the head area, then spiral hackle toward tail. If you selected a tapered hackle, it should look something like this. Secure the last turn of hackle at the tail area with wire rib.*

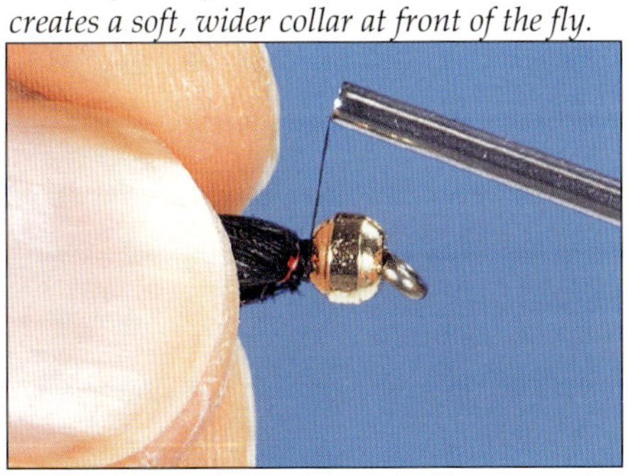

10. *Take extra turn of rib at head, secure with thread, trim off end. Pull all hackle fibers back away from bead and complete a small tapered thread head. Lacquer, and you are finished!*

11. *Finished Woolly Bugger. Note proportions, hackle placement, and taper. Proceed onto the Misfit™ and then the next four patterns, all of which are dry flies that have palmered hackle.*

Misfit™
1. Slide cone onto hook and cover with thread. Pictured is a thick marabou plume with center stem cut out. When pulled together, the ends are of equal length. This fly has a two-color tail.

2. Select and prepare a white plume and match the tips. Both plumes have been matched together and dampened for better control. This tail should extend beyond the hook as far as the hook is long.

3. Thread is positioned at exact tie-in location. Hold plume in right hand to measure tail length. Grasp marabou with left hand and pinch at the tie-in area. Secure with a few tight turns of thread.

4. Marabou is secure at tie-in location and must be tied down along the top of the hook back to the tail area. Hold the tail in place with your left fingers as you wrap the thread.

5. Tail and underbody foundation are in place. Note thread placement—a few at head area, spirals to tail, and several tight wraps to keep it in place. Butt ends have been trimmed closely and at an angle.

6. Tie in a few strands of Mirror Flash or holographic Flashabou along one side of the tail and fold the ends along the other side of the tail. This saves making two cuts and two tie ins. Trim even with tail.

7. Secure fine silver wire and Crystal Chenille. Rub off ends of chenille to avoid building bulk. Wrap thread to front of body area as pictured. Wrap chenille to this point, stroking it back after each wrap, and tie off.

8. Arrow points to the trim point of this saddle hackle. It includes some fluff as we want this hackle to be "soft" so it soaks water like a sponge and flows back along the abdomen.

9. Hackle ends have been stripped for 1/16-inch. Hackle is tied on its side. We are looking at the dull side. When wrapped, this side faces back. Take three turns at the tie-in area and palmer back to tail area.

10. Once hackle is at tail area, secure with fine wire and spiral wire to front of body. Take two turns in place at this area, tie off with the thread, and trim excess. Hackle is complete.

11. Notice the space behind the cone. If fly looks like this, it is perfect, and there is plenty of space for the chenille head. Twist off the end of chenille to its core and tie in as pictured. Wrap until the area behind cone is full.

12. Finished Misfit™.

Tying Dries

Everyone loves to fish dry flies. The visuals of a trout capturing an insect at the water's surface captures the imagination of every person who is witness to it. This almost mesmerizing sight converts more anglers to fly fishing than any other. There *is* magic in the ring of the rise.

Fishing a fly with a natural (dead) drift on or at the water's surface is easier than fishing a nymph in the same manner below the surface, simply because you can see the fly, the currents and drift of the fly, and the fish—often before and when it takes the fly! What could be better? Not much in the world of the trout angler.

Tying dry flies is equally rewarding, but in contrast to fishing them, dry flies are generally more difficult to tie than nymphs. Dry flies should usually be more exact than nymphs, and most important, they must float! For best results, they should balance and be properly proportioned. These traits are accomplished by using light wire dry fly hooks, water repellent materials or those with floating qualities, and stiff hackle. Applying fly floatant helps to keep them from absorbing water. Keeping a mostly tensionless, drag-free drift on the water's surface also helps to keep them floating. A tight line and leader pulls the fly under the water's surface.

Wing and hackle style mostly determine the type or style dry fly you tie. There are many different types of dry flies; hence there isn't any one specific guide to proportions. Standard upright and divided wing and hackled dry flies do, however, usually fall into the proportion guide illustrated here. There are many styles of various wings and hackle, or lack of one or the other, or both. Wing styles include spent, clump, thorax, down hair, upright and divided hair, down feather, down hackle tip, upright hackle tip, divided feather, quill, various parachute, and even double wings. It is beyond the confines of this book to detail all of them. Tyers interested in learning more about dry fly wings, hackle, proportions, tying with hair, and how best to represent the adult stages of surface insects should consider my book, *Tying Dry Flies.*

Like all styles of flies, dry flies can be tied as exact or nearly exact imitations of specific insects, or as imaginative, unorthodox attractors. Most, however, are species specific, general, impressionist, or multi-imitators.

In this section we learn to tie upright-and-divided feather wings with the March Brown, upright-and-divided hackle-tip wings with the Adams, a down hair wing with the Elk Hair, a double downwing with the Hot Butt® Caddis, and a parachute wing with the Adams Parachute. All are on the top 10 list and provide an excellent beginning in understanding the nuances of imitating surface insects, or tying dry flies.

Dry Fly Proportions (900BL)

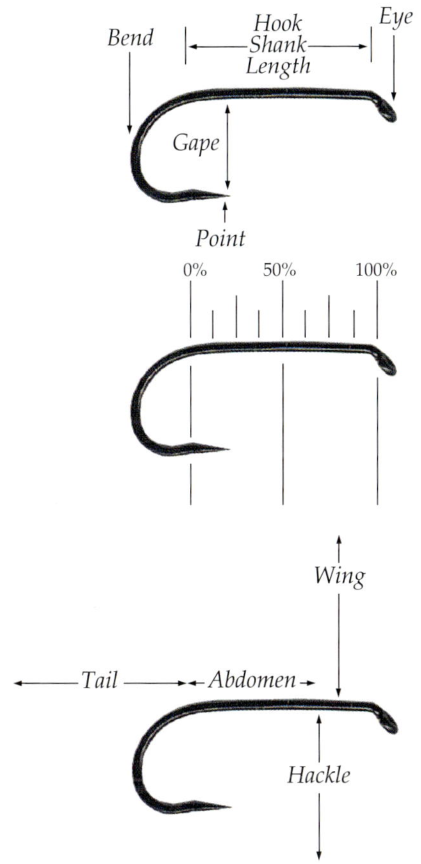

Upright wings are generally as tall as the hook shank is long and are positioned at the 65-85% mark. Downwing, spent wing, etc. may be different.
Tail length is the same as the hook shank length.
Abdomens cover the hook to the 60-70% mark.
Hackle begins at the 60-70% mark and continues to the thread head area. It extends about 1½ times the width of the hook gape.

Griffith Gnat *(George Griffith)*
Hook: 900BL, sizes 12-22
Thread: Black or to match abdomen
Rib: Thread or fine gold wire
Abdomen: Peacock or other to suit
Hackle: Grizzly or other to suit, palmered

X Caddis *(Craig Mathews and John Juracek)*
Hook: 900BL, sizes 12-18
Thread: Tan or to match abdomen
Tail/Shuck: Gold Z-Lon (substitute Antron Yarn) or to match abdomen dubbing—represents shuck
Abdomen: Golden tan, golden olive, Antron or other dubbing, color to suit
Wing: Light mottled deer hair, tied downwing style, butt ends clipped short

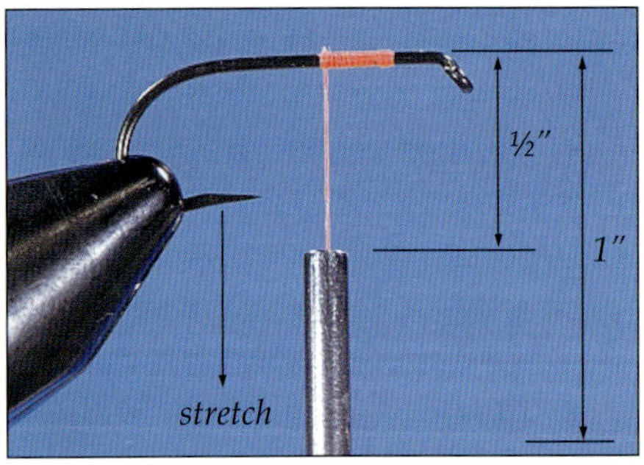

Depending on the type thread and the distance it extends beyond the bobbin tip, it may stretch to almost double its length. A half-inch of thread can be stretched to an inch. Ease into tight tie downs, stretching the thread slowly. Due to the impact force, a quick or hard pull breaks it every time.

Chapter 16

Griffith Gnat and X Caddis

The flies in this and the next chapter have been selected for their fish and angler appeal, and for tying technique. They demonstrate how techniques and materials build upon one another. Eventually, only an occasional additional material or tying technique might be needed to construct a new pattern.

The Griffith Gnat is one of the simplest dry flies, consisting of three parts—rib, abdomen, and hackle. It is a refresher course in working with peacock and a palmered hackle, and should be an easy tie. Remember, the colors and materials can be adjusted to your needs. Peacock is available in dyed colors, and a blue dun hackle combination is quite striking. Another possibility is to use ostrich or dubbing in lieu of peacock.

Anglers usually fish Griffith Gnats for adult midge. Sometimes larger sizes 12 and 14 are called upon to represent "rafts" of midges that congregate for various reasons, or are brought together by wind and current. Griffith Gnats also double for gnats, black flies, and other tiny flies if fish are not too picky.

The X Caddis, designed by Craig Mathews and John Juracek, is another simple three-part dry fly—tail (shuck), abdomen, and wing. Hatching, emerging, or crippled adults may have a "shuck" attached to them, which is imitated by a "tail." The actual shuck is the nymphal case from which the adult insect exits from at or near the water's surface. Some insects become trapped in their cases and cannot escape, becoming stillborns or cripples, which are easy targets for fish. Shucks are often represented with Z-Lon, Antron yarn, and/or CDC.

The X Caddis is a quick study in tying a hair downwing. Most hair downwings are tied the same—simply secure the material onto the top of the hook. Beginning tyers often break 6/0 thread when working with hair. Once you have the feel for and know how far the thread stretches and how much pressure you can exert against "soft" and "hard" materials like hair, 6/0 should be ample. Size 6/0 is less likely to break if you slowly ease into extreme pressure. Anglers know that 5x monofilament breaks easily with a quick jerk, but, when stretched slowly, it is much harder to break. Tying thread works under the same principle.

Sometimes the butt ends of hair downwings are trimmed close and completely covered with thread, or left extended and trimmed even with the hook eye, as with the X Caddis, Hot Butt®, and the Elk Hair Caddis.

Griffith Gnat
1. Install fine wire (for rib) and peacock abdomen. Prepare and tie grizzly hackle at back of thread head area. You have already accomplished and practiced these techniques.

2. Wrap palmered hackle, dull side to back, taking two turns at front before beginning spiral wrap to back. Once at the rear of the abdomen, secure with fine wire and continue wire through abdomen, tying off at head.

3. Trim ends of wire and hackle tip at back of abdomen and complete tiny thread head. Let's take a slight detour from hackle and construct an X Caddis, installing a hair downwing. In the next chapter we put the two flies together.

X Caddis
1. Tie in an Antron tail (represents "shuck") and add sparse, thin dubbing. It does not take too much, especially if I had used light-colored thread. I have used black for visibility.

2. Wrap dubbing to the 85 percent mark, forming a small diameter taper. Do not infringe on the thread head area. Position the thread as pictured. Note thread base where wing will be tied.

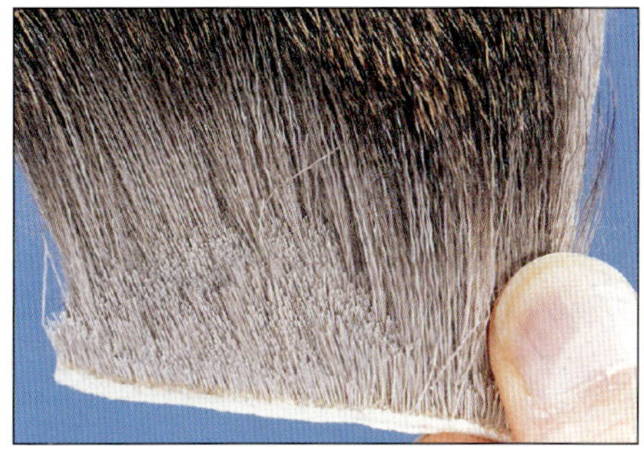

3. Pictured is a piece of deer hair that is partly used. Grasp a couple dozen fibers with one hand and trim them off close to the skin. Deer is softer than elk and, unless it is very stiff, flares easily.

4. This bunch of deer hair becomes the wing. Hold by tips, pull out any short or fuzzy fibers, then stack the tips until they are even. A large diameter stacker is best because material can spread out and fall inside, speeding up the process.

5. Place the stacked bunch of deer over the top of the abdomen at the head area as pictured. Wing should extend about to the bend of hook, but length, like anything on a fly, can be whatever you want it to be.

6. Holding wing in position, grasp hair with left thumb and first finger. Pinch and secure with one to two turns of thread at the position pictured. Hair should flare as seen in next photo.

7. Back fingers off slightly as pictured, still pinching wing in place, and secure with six well-placed tight turns of thread. Ease into thread pressure; it is easy to break 6/0 on hair. Note position of tie down.

8. Wing is secure. Sweep butts back away from hook eye, take a few turns of thread under or in front of wing butts, and tie off the thread. Grasp all butt ends, pull forward evenly over hook eye and trim even with eye.

9. Finished X Caddis. The next pattern is an Elk Hair Caddis. It is a combination of the Griffith Gnat and X Caddis. A simple, four-part dry fly (abdomen, rib, palmered hackle, and wing), it is favored by many anglers.

Elk Hair Caddis, Original *(Al Troth)*
Hook: 900BL, sizes 10-20
Thread: Tan
Rib: Fine gold wire
Abdomen: Hare's ear dubbing
Hackle: Medium to dark brown, palmered
Wing: Cream or bleached

Hot Butt® Caddis *(Randall Kaufmann)*
Hook: 900BL, sizes 12-18
Thread: Black or to match body
Butt: Fluorescent red Glo-Bug yarn or other color to suit
Rib: Fine gold or copper wire
Abdomen: Peacock herl or dubbing to suit
Hackle: Grizzly, or other color to suit, palmered through body
Underwing: Natural gray CDC tied full and thick
Overwing: Natural gray elk tied sparse
Note: I like a peacock abdomen and fluorescent Antron colors—the brighter the better.

Underview of Hot Butt® Caddis. The fly should have lots *of dense CDC in the wing and* very little *elk. Spread the CDC wide. This is an important aspect of the fly. Do not treat CDC with floatant. Dry slowly, use a drying agent, or change flies. Good luck!*

Chapter 17

Elk Hair Caddis and Hot Butt® Caddis

Combine the palmered hackle of the Griffith Gnat with the deer hair wing of the X Caddis, and you have an Elk Hair Caddis, which is one of the world's favorite flies. Designed by ace Montana tyer Al Troth, it floats high, is easy to see, and easy to tie. No new techniques, just a combination of those previously practiced that form a terrific fly. We are tying a relatively sparse version for slower water. It is all too easy for beginning tyers to overdress flies, so I stress less is better, and a sparse wing is easier to secure tightly than a heavy wing.

The Hot Butt® Caddis is similar to the Elk Hair Caddis with the addition of a fluorescent butt and a CDC underwing, making it a little more difficult, six-part pattern. I have experimented with fluorescent-colored flies for years, and they often seem to be favored by trout. I rely more and more on bright neon patterns, and many of my favorite Stimulators and other dry flies and nymphs are fluorescent.

Joe Burke, my friend from Jackson, Wyoming, was fishing some of his fluorescent butt nymphs with exceptional results while I went nearly fishless. There was something to those fluorescent tails! It reminded me of the time Dennis Black, of Umpqua Feather Merchants fame, gave me a lesson in New Zealand with flies tied with CDC long before anyone in this country used them. CDC is unique in that individual barbules spiral *around* the feather stem instead of extending at right angles from the sides, preventing them from clinging or marrying together. This creates space and air pockets, which generate a unique light pattern and water imprint.

I mention this as a brief insight as how various pattern and material ideas from others meld together, helping form new patterns. There are few "new" fly inventions—mostly the reshaping of older ones that incorporate new materials and ideas from others. By combining favorite materials into effective patterns, my fly box always has a "new" appearance, which trout seem to appreciate.

Dense CDC puffs must be used for the Hot Butt® Caddis wing, which should be *thick* and spread *wide*. The elk hair overwing is actually "topping." It helps control the shape and somewhat confines the whimsical CDC puff; it should be *sparse*. Do not treat CDC with floatant because it matts the barbules reducing its floating qualities and uniqueness. Blow it dry, use a drying agent, or tie on a fresh fly. Peacock and fluorescent yellow, orange, green, red, and cerise are some of my favorite colors, but use your imagination.

Elk Hair Caddis

1. Tie in the fine wire rib, construct the abdomen, and tie the palmered hackle in place, dull side facing you, angled off the opposite side of the hook as pictured.

2. Complete the palmered hackle. Bring the rib forward, and trim off the ends of the rib and hackle. By now, these techniques should be second nature—or almost so!

3. Prepare and tie in an elk hair wing (same technique as for the X Caddis), and the Elk Hair Caddis is finished. The trick is to use stiff elk for the wing and to tie it down securely.

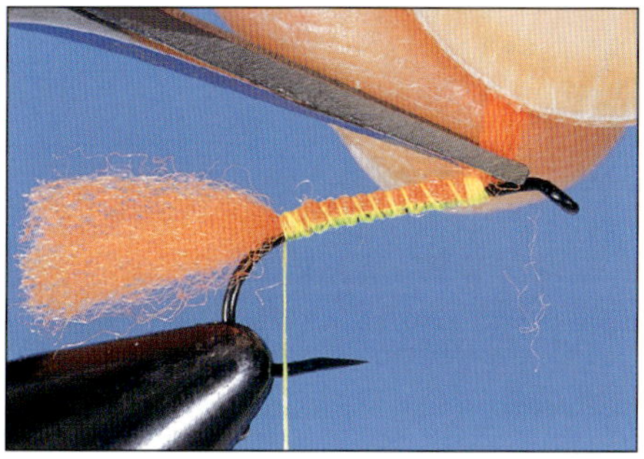

Hot Butt® Caddis

1. Secure Glo-Bug yarn tail, tapering the end up front and trimming the butt close and even. Then, tie in the fine wire and dub the body, keeping it slender—see next photo.

2. Complete a palmered hackle, which should be easy by now! It should be about 1½ times as wide as the hook gape. Take two turns at the front, then spiral hackle back to tail, secure with wire rib, and finish rib.

3. Pictured is a CDC puff. Note shape and density. Demonstration fly is a size 10, so we need at least two feathers. CDC wing should create the prominent aspect of the fly. See bottom photo on page 66.

4. Pinch CDC with left hand and tie in place. It is standing upright here because of hackle. It will settle, and elk topping helps keep it down. Notice the space left toward eye to complete wing and thread head (see photo 5).

5. Dampen CDC for better control and a clearer view. Trim butt ends as closely as possible. One can only trim materials as close as the scissor blade is thin. Use a thin, fine point, sharp scissors.

6. Select a few elk fibers (about 25 percent of what an Elk Hair caddis would require) for the topping, which helps hold the CDC in place. Even the hair tips in a stacker, which makes a much nicer-looking fly.

7. Measure wing length with hair in right hand. Change hands, pinch in place, and secure. The elk is a "harder" hair than deer, so it is easier to break thread. Make the first turn semi-tight, slowly cinch, then go for it.

8. As you increase thread pressure on hair, it compacts together and compresses against the undersurface of thread. Secure as tightly as possible. Lift butts of hair and finish head and knots underneath and in front of wing.

9. Trim hair butts as pictured in photo 8 and lacquer. Notice proportions. Hackle is 1½ times hook gape. Wing extends to hook bend. Spread CDC until it is very wide.

March Brown
Hook: 900BL, sizes 10-16
Thread: Brown
Wing: Mallard flank dyed woodduck
Tail: Brown hackle fibers
Rib: Copper wire or yellow floss, twisted tightly
Abdomen: Brown pheasant tail fibers or dubbing
Hackle: Furnace or brown

Light Cahill (Dan Cahill)
Hook: 900BL, sizes 12-20
Thread: Pale yellow
Wing: Mallard flank dyed woodduck, upright and divided (woodduck flank was originally used)
Tail: Light ginger or cream hackle fibers
Abdomen: Light ginger or cream Antron or Superfine
Hackle: Light ginger or cream

Chapter 18

March Brown and Light Cahill

The March Brown is usually the first big (sizes 10-16, usually 12-14) mayfly to appear in the spring. Depending on elevation and weather, mid-March to early May is usually the peak time. March Browns can be slow to get off the water, so trout have ample opportunity to sip them at the surface, which provides excellent action.

Light Cahills are an early summer to midsummer mayfly that are more prevalent in the eastern US, less so in the Western U.S. However, the Light Cahill color represents many cream-colored mayflies, especially PMDs and Sulphurs, so the fly enjoys universal popularity in sizes 12-18.

From the standpoint of tying technique, both patterns are the same; only the colors change. Many other standard upwing dry flies are of similar design, including the Dark Cahill, Quill Gordon, and Hendrickson. These patterns, developed from the Catskill school of tying, were originally tied to be sparse and delicate, with wings a bit longer than normal, and cocked slightly forward. These instructions are middle of the road, not too sparse, not too heavy.

The techniques to master with the March Brown are standard dry fly hackle and upright and divided feather wings. Upright and divided wings are best tied in first. Once they are properly positioned and proportioned, they are used as a guide, and everything else falls into place perfectly. Follow the photos and captions carefully and this wing is a snap. Note that after wings have been tied in place, their angle can be adjusted with hackle placement, and at any time with your fingers. Installing dry fly hackle is easy once you understand how to select, wrap, and tie it off. *Use the best possible hackle*. Don't settle for second best—it only frustrates you.

March Brown
1. Select a mallard flank feather dyed woodduck color with even tips. These come in many shapes and sizes and curvatures. Use two feathers for larger flies, or two feathers if they are smaller and less select.

2. Remove the shorter fibers that do not reach the length of the others from the base of the feather. Trim out the center stem. The center stem controls many fibers, and we do not want it in the finished wing.

3. Pull all fibers together, grasping them at the tie-in length, which is about the length of the hook shank—wings are as tall as the hook shank is long. Notice that the fibers are all the same length, extending to the same point.

4. Secure on top of the hook at the 80 percent mark. Trim the ends at a sharp angle but not too close to the thread tie down. There must be enough mass to catch the thread when dividing the wings (see photos 7 and 8).

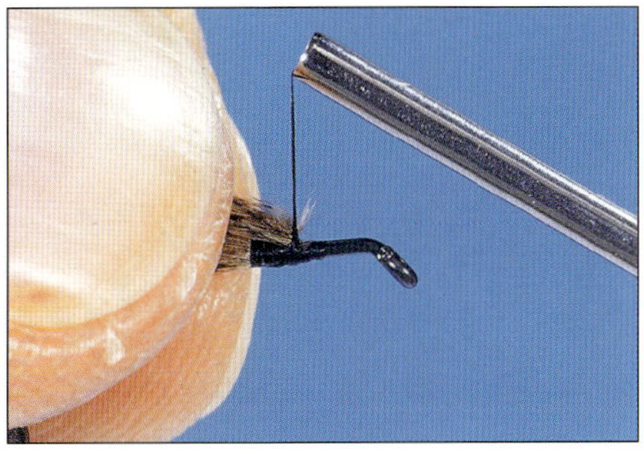

5. Wrap a few turns of thread immediately in front of the wing, standing it up. Only if thread is wrapped against *the wing does it help stand it upright.*

6. Wing "clump" is upright. Thread is at front of wing. With your fingers, divide the wing in half. Next, X thread between wing halves into wing butts. It helps to hold one side of the wing with your fingers.

7. Bring thread back between divided wings, hooking into butt ends on opposite side of hook, which helps separate them. Then take a turn of thread around entire hook behind wing.

8. Catch the thread in the wing butts on the near side of the hook and bring between the wings toward the front. Take two thread turns in front of wing to lock everything in place.

9. The longest and stiffest hackle fibers are found along the middle sides of a neck. Referred to as spade hackle, they make excellent tails. Even fiber tips by first arranging them at 45-degree angle from stem, then pull off. *See page 71, photo 9.*

10. Beginning at wing butt, secure hackle fiber tail onto top of hook. Holding the fibers together and slightly above the hook with your left hand, wrap the thread to the hook bend, tying down the tail. Position the thread as pictured.

11. Beginning behind the wing, tie in a small bunch of ringneck tail fibers by the tips and a three- to four-inch section of fine copper wire. Secure ringneck and rib along top of, and to back of, hook to tail area. Reposition thread as pictured.

12. Wrap the ringneck fibers forward, maintaining constant tension, forming the abdomen. Then wrap wire rib. Tie both off just short of wing. (You need space for hackle.) Notice the smooth and level underbody in photo 11.

13. Select a hackle (this is furnace, which is brown with a black center) and tie in so the shiny side faces you. Begin wrapping the hackle back to front, shiny side toward the back, each turn immediately in front of last.

14. Take perhaps three turns behind the wing, with two turns in front, or 4/2, 4/3, 3/3, or 2/2 for a sparse hackle. A heavy hackle might be 6/5 or 7/5. Secure hackle stem tightly against the last turn of hackle. Trim ends back away from the hook eye as far as possible.

15. By doing this, trimmed ends do not protrude into or over hook eye (photo 14). A neat head is easily constructed. Try tying a brown and a grizzly hackle together, called an "Adams hackle." This makes a beautiful March Brown. Perhaps use yellow floss for rib.

Blue-Winged Olive

Hook: *900BL, sizes 14-20*
Thread: *Olive*
Wing: *Dark blue dun hen hackle tips, upright and divided*
Tail: *Dark blue dun hackle fibers*
Abdomen: *Medium olive to brown olive Superfine or Antron*
Hackle: *Dark blue dun*

Pale Evening Dun *(Charles Fox and Ray Bergman)*

Hook: *900BL, sizes 14-18*
Thread: *Cream or tan*
Wing: *Blue dun hackle tips, upright and divided*
Tail: *Light ginger or blue dun hackle fibers*
Abdomen: *Cream with yellow-orange highlights*
Hackle: *Ginger*

Chapter 19

Blue-Winged Olive and Pale Evening Dun

Baetis, commonly called Blue-Winged Olives, inhabit almost every clean waterway in North America, from sea level to over 12,000 feet. These tiny (size 16 and smaller) olive to olive-brown-colored treasures can be found hatching someplace on the continent every day of the year. Some of the most spectacular hatches occur during fall, winter, and spring between storms and cold weather days, especially on cool, damp, overcast days. Pick a blue ribbon stream in the West. It is certain to have a big population of *baetis*. Of the upright-and-divided wing mayfly patterns, Blue-Winged Olives are perhaps the most popular of the day, and belong in every fly box.

The Blue-Winged Olive provides practice installing standard-style dry fly hackle and demonstrates upright-and-divided hackle-tip wings. This fly, as shown here, is designed for slower water, incorporating a hackle fiber tail and light or sparse hackle.

The Pale Evening Dun is a similarly-constructed fly. It represents *Ephemerella dorothea,* an early summer evening hatch that affords midwestern and eastern anglers excellent surface action. The nondescript colors of this fly make it of value everywhere.

Hen hackles are usually best for hackle tip wings because they offer a broad silhouette, are less expensive, and are easy to handle, but rooster neck hackle tips are sometimes used.

By tying the wing first, all tie-downs are covered and the stage is set for proportions. When the wing butts are tied down toward the hook eye, it is easy to stand the wing up by placing one turn of thread behind the wing and between the stem and hackle fibers, but don't crowd the head area. All in all, this wing method is the most efficient.

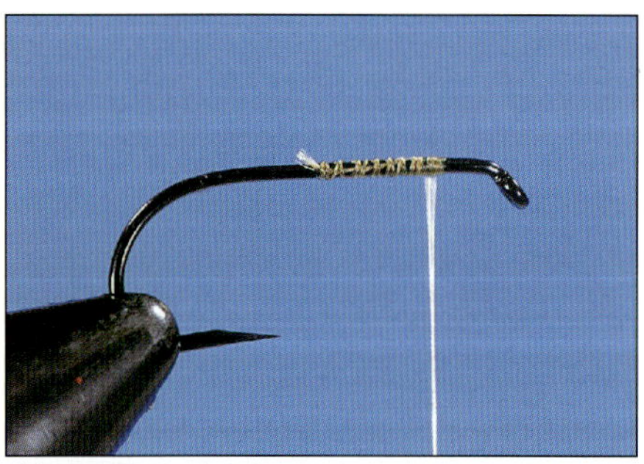

Blue-Winged Olive
1. Wrap base of thread. Position at 80-percent mark. The first securing turn of thread over hackle tip wing occurs here. Succeeding turns of securing thread are wrapped back *from this position.*

2. Select two like-sized hen or rooster hackle tips, even tips and face shiny sides together—convex sides together. Hackle flares away from each other, creating a **V***. These become "upright and divided hackle-tip wings."*

3. Hold wing as pictured to measure length, which should be the length of hook shank. Keeping the wing in place, change hands, pinch, and tie in at the 80-percent mark. This is the technique for tying downwings (Mosquito, Timberline Emerger).

4. Bring thread up between your fingers and down, locking wing in place. Back fingers away slightly and secure by wrapping follow-up turns of thread toward the hook eye, then back over original tie-in position, keeping wing positioned on top of hook.

5. Wing is properly placed. Note length and thread placement. Trim butt ends as closely as possible to avoid infringing on hook eye.

6. Tilt wing forward and place a wrap of thread behind and tightly against feather stems, tying down fibers. One turn should stand wing upright. Secure, wrapping back with two to three more turns. Wing can be manipulated at any time.

7. Wing is held in place by thread in front and behind. Grasp all fibers slanting back, tie down onto top of hook, forming level foundation, but trim any excess that might be too long.

8. Errant wing fibers have been tied down and any longer ends trimmed. Notice the even and slightly tapered underbody. Thread is positioned where hackle fiber tail should be tied onto the hook and is melded into taper.

9. Spade hackle is found along the side of rooster necks. It is the stiffest and longest. Leave remaining hackles on neck so you can select another tail from the same hackle later.

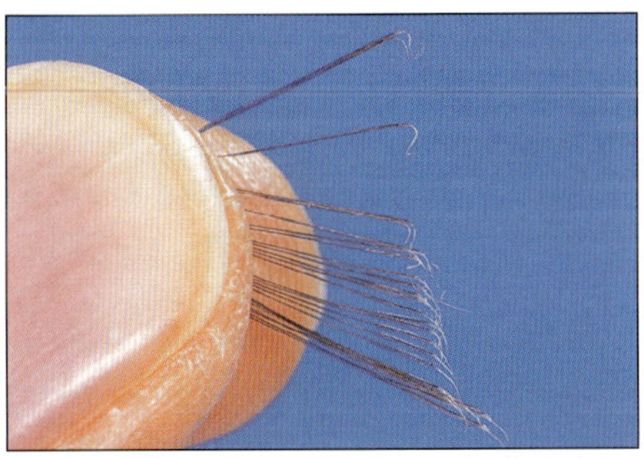

10. Hackle fibers in close proximity along hackle stem are the same length. Even the fiber tips by arranging them 45 degrees from stem before removing them. If you did it right, the butt ends of fibers should be even. If they are even, tips are also even.

11. Bunch and grasp the butts with right hand. They should be even. Fifteen to 18 fibers is about right for a size 10 hook, 10-12 or so for a size 14. Hold fibers against hook to measure distance.

12. Tail fibers are tied onto hook behind wing. Note the butts extend past wing. Trim cleanly and tie fibers down along top of hook back to tail area, helping form a slightly tapered foundation. See next photo.

13. Dub abdomen, keep slim (photo 14). I sometimes like to dub a little in front of wing, creating a soft, even, and continuous base for hackle to adjust into.

14. Tie hackle on edge, angled away, looking at the shiny side. Wrap hackle so the shiny side faces back toward the tail. Begin at rear-most area that hackle will occupy and wrap to front, each turn adjacent to last.

15. Wrap hackle forward until tight against back of wing. Then take a tight turn of hackle tightly against front of wing, helping to secure wing in upright position. Leave no gaps between turns/wraps of hackle. This is a sparse tie. Finish and lacquer head.

Adams *(Leonard Halladay, 1922)*
Hook: 900BL, sizes 10-20
Thread: Gray
Wing: Grizzly hen hackle fibers, upright and divided
Tail: Grizzly, grizzly and brown hackle fibers, or moose
Abdomen: Muskrat or gray Antron or Superfine
Hackle: Grizzly and brown mix (Adams)

UFM

Adams Parachute
Hook: 900BL, sizes 12-18
Thread: Gray
Wing: White calf, parachute style
Tail: Grizzly and brown hackle fibers
Abdomen: Muskrat or gray Antron or Superfine
Hackle: Grizzly and brown, parachute style
Note: Adams parachute wings can be any color. Tie what is easiest for you to see. Under certain sun and glare conditions, black, fluorescent orange, or fluorescent red may be best. I have them all.

Chapter 20

Adams and Adams Parachute

If faced with the difficult choice of fishing only one standard-style dry fly, many anglers would select the Adams. It can represent many foods, including gnats, mosquitoes, deer flies, craneflies, and various species of mayflies, the most popular being the gray drake, black drake, *callibaetis,* and even *baetis* in small sizes.

Unlike the previously constructed Blue-Winged Olive, the Adams demonstrated is designed for fast water, incorporating a stiff, better floating moose tail and heavy hackle for better flotation. Otherwise, tying techniques are the same as for the Blue-Winged Olive.

Anglers love the Parachute Adams, the parachute version of the basic Adams. Its gray-brown color combination entices fish during many hatches. As an added bonus, it floats well and the white wing is relatively easy to follow, even in fast, choppy water. Fish see it, too, because they eat lots of Parachute Adams.

According to Ray Bergman, the original parachute was called a gyrofly and was introduced by A. C. Mills, Sr., through the former sporting goods dealer William Mills and Son of New York. The flies were under patent by William Avery Brush of Detroit in 1934 and were tied on a hook made especially for the purpose, or hackle was wrapped around its own stem. Like many other original fly patterns, parachutes have evolved and have been vastly improved upon and adapted to many other styles of tying.

Today parachute-style flies have a well-deserved permanent place in the angler's fly box. Their unique horizontal hackle construction usually allows them to flutter down lightly and land in an upright position without sacrificing floating qualities. They float flush in the surface film, emphasizing body size, color, and wing silhouette. They represent a variety of food sources, including stoneflies, mayflies, caddisflies, and craneflies. Parachutes also possibly represent spent mayfly spinners and drowned and otherwise disabled duns.

When first seen, the technique for tying parachute hackle always elicits comments like, "so that's how it's done," or "isn't that tricky?" Parachute hackle is usually wound horizontally around a single upright wing. Each succeeding wrap of hackle is taken *under* the last wrap. If the wing base is not rigid, you must hold onto the wing when you wrap the hackle, or the wing bends and the hackle slips off over the top of the wing.

Once mastered, parachute flies will become a favorite at the tying bench and on the water. Adapt the materials and colors to specific mayflies, such as *baetis, callibaetis,* PMDs, etc. See the pattern directory.

Adams

1. Two like-sized hen hackle tips have been prepared by placing the shiny (convex) sides facing each other and by evening the tips.

2. The hackle tip stem has been stripped of fibers, but not all the way to where the wing will be tied onto the hook. Some fibers are needed to join in the tie-down process; otherwise the bare stem twists and turns, and doesn't tie in as securely.

3. Pinch at tie-in length, which is length of hook shank they are to be tied onto, and secure in place with your scissors—measure wing length against hook length. These are a shade on the long side, but we'll go with it.

4. Trim butt ends off, away from hook eye. Lift wings forward by their tips, allowing shorter fibers to fall free. Place thread tightly against wing stem, locking wing upright.

5. Wing and moose tail are in place. Use a stacker to even moose hair fibers for tail. Note proportions, smooth underbody, and space in front of wing for hackle and head. Gray thread blends with overall fly color. Tail is flared for balance.

6. Install dubbed body as pictured—thin taper! Wrap a little dubbing in front of wing to keep everything smooth and even and to provide a soft base for hackle. Position the thread as pictured. Hackle will cover space from here to thread head.

7. Tie in two hackles. Once mastered, wrapping two at a time is efficient and creates a better hackle. Tips must be even, stems must be aligned and tied in as one. If possible, use your fingers to wrap hackle—it is difficult with pliers. Try it!

8. To begin, pull stems and hold together close to base of hook shank. Do not allow them to spread apart. Wrap tightly against wing, pull wing back, and wrap against front of wing, positioning wing upright. If you can't wrap two hackles, do one at a time.

9. Three turns in back, two in front with double hackle makes a fast water fly (10 turns of hackle.) Grasp tips closely as possible and place turn of thread around hackle stem, being careful not to tie down hackle fibers.

10. Trim tips cleanly and closely. This is best done by sweeping all fibers between your right thumb and first finger. Pull back away from hook eye. Place fine-point scissors against hook shank and head and trim. Finish thread head and lacquer.

11. Top view of finished Adams. Tail extends straight from hook shank and is as long as hook shank. Note that this fly is heavily dressed, meaning it has lots of hackle, and is designed for heavy water.

12. Front view shows the V wings, diameter of hackle, and wing length in relation to hackle. Upright wings generally extend to about the height (length) of hackle or slightly longer. Hackle is usually as high as the hook shank is long. These wings are a bit too tall.

Adams Parachute

1. Position thread at the 80-percent mark. Select and stack some calf body hair, which is straighter than calf tail. Take a loose wrap up-between-the-fingers and cinch it tightly.

2. Wrap about 10 turns of thread back toward tail. Wing should be secured as shown. Trim butts at an angle, but not too closely or it is difficult to catch butts with thread when you stand the wing up.

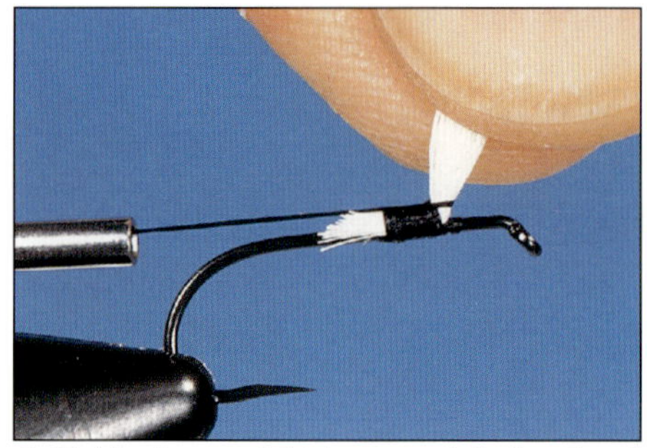

3. Hold wing upright in your right hand and bring thread around horizontally in front of wing. Wrap must be placed high on wing at position shown, or it will not stand up. If thread is placed too high, wing slips away.

4. Once thread is wrapped in front of wing, it is brought around and through butts, as seen here. Next, take three to four standard wraps that lock stand-up horizontal wrap in place.

5. Select a few hackle fibers and tie them in just behind wing butts. Tie tail down back to tail area and trim off butt ends. Resulting underbody should look like this.

6. Dub and wrap body. Front part of abdomen can be wrapped now or after hackle is in place. A touch-up turn or two of thinly-applied dubbing is often needed to cover up any noticeable thread wraps. Gray thread would not be noticeable through dubbing.

7. Tie hackle in behind wing flat, not on edge, with shiny side facing up. Stripped hackle stems that are tied down can extend in front of wing where they can also be tied down. If needed, wrap more dubbing in front of wing.

8. Parachute-style hackle is wrapped horizontally around base of wing. A thread base may be placed around base of wing. Wrap first turn of hackle at bottom of wing. Succeeding turns are placed below last wrap, pushing preceding turn of hackle up the wing.

9. To tie off hackle, sweep all fibers back between your fingers. Attach hackle pliers to hackle, freeing your right hand. Tie off hackle at rear of thread head area. Notice bare hook behind hook eye. Do not crowd head area.

10. Sweep hackle fibers out of the way. Place scissors against hook shank and make one clean cut. Pull hackle back with your left hand and finish thread head. Rearrange hackle and wing into desired position.

11. Finished Parachute Adams.

12. Underside view. Note symmetrical hackle and tapered abdomen.

Pattern Directory

If you have progressed diligently through this manual and mastered all 28 demonstration flies, you should be able to tie any of the 36 patterns presented here. They represent many productive and frequently fished flies throughout the fishing world. You should be able to "read" these "patterns", understand the tying sequence and techniques, and visualize the finished fly. If you have trouble remembering how to accomplish a specific technique or "part" of a pattern, check the index and refresh yourself.

Dressings are given in the order in which they are *tied onto* the hook. This is not necessarily the order in which they are *wrapped around* the hook or *manipulated* into final position. As an example, the wire rib for the Woolly Bugger is tied *onto* the hook before the hackle, but is not wrapped *around* the hook into final position until the hackle has been wrapped into place. Good feather bending…and fill those fly boxes!

Umpqua Feather Merchants

Bird's Nest *(Cal Bird)*

Hook: *3761BL, sizes 8-18, weighted*
Thread: *Tan or to match body*
Tail: *Mallard or teal flank dyed woodduck or to match abdomen*
Rib: *Medium copper wire*
Abdomen: *Grayish tan Australian opossum or other to suit*
Legs: *Same as tail, tied full and extending to barb or slightly beyond*
Thorax: *Same as abdomen*
Note: Favorite California pattern

Blair Alexander

Blair's Baiter™ *(Blair Alexander, 1995)*

Hook: *9394 size 6, or 5263 size 4*
Thread: *Olive*
Abdomen: *Pearl diamond braid*
Wing: *Chartreuse over white bucktail highlighted with 2 strands each side of pearl Flashabou*
Head: *Build up with thread*
Eyes: *Chartreuse 3-D moulded, size 7/32, epoxy over*
Note: Represents candlefish, (sand lance) and fished for silver salmon and cutthroat.

Borealis™ *(Randall Kaufmann, 1996)*

Hook: *7999, sizes 2-4*
Head: *Gold cone*
Thread: *Fluorescent red*
Tail: *Hot orange over white marabou with Krystal Flash, Holographic Fly Flash, and Flashabou highlights (pearl, fl. orange, gold, silver)*
Wing: *Cerise over white marabou staggered along hook with Krystal Flash and Flashabou highlights*
Hackle: *Cerise, long and soft*

UFM

Brassie *(Gene Lynch)*

Hook: *5262, 3761BL, or 900BL, sizes 12-18*
Thread: *Black*
Abdomen: *Copper wire*
Head: *Black dubbing*
Notes: Simple, unassuming pattern that has accounted for many excellent catches. Especially popular in Colorado.

Highland Flies

Bright Roe

Hook: *7999, sizes 2-8, weight to suit*
Thread: *Fluorescent red, green, or orange to match abdomen*
Wing: *Glo-Bug yarn, color of choice, with holographic and pearlescent reflective highlights*
Rib: *Holographic tinsel*
Abdomen: *Fluorescent red, green, or orange chenille*
Note: Trout, char, Dollies, steelhead, and Pacific salmon.

UFM

California Mosquito

Hook: *900BL, sizes 12-18*
Thread: *Black*
Tail: *Grizzly hackle fibers*
Rib: *White thread*
Abdomen: *Black floss*
Wing: *Grizzly hackle tips, downwing style*
Hackle: *Grizzly*
Note: Most effective mosquito pattern; especially popular in western America. For wing tie in, see page 70.

UFM

Cased Caddis, Bead Head

Hook: *200RBL or 5263BL, sizes 6-12, weighted*
Head: *Black, gold, or copper bead*
Thread: *Black*
Rib: *Copper or gold wire*
Abdomen: *Peacock herl, grizzly, brown or furnace hackle, palmered and clipped at angle; represents rock case*
Thorax: *Yellow to pinkish-orange Hare-Tron or other color to suit*
Head: *Black HareTron, picked out*

Cate's Turkey *(Jerry Cate)*

Hook: *900BL, sizes 14-18*
Thread: *Black*
Tail: *Mallard dyed woodduck*
Rib: *Gold wire—entire abdomen-head*
Abdomen: *Turkey tail*
Thorax: *Peacock*
Legs: *Mallard dyed woodduck, beard*
Notes: Developed at Davis Lake, Oregon, in the late 1960s by Jerry Cate, who was a regular at the lake. He fished this fly very slowly. Represents Callibaetis*. It is good everywhere* Callibaetis *are found.*

UFM

Chironomid Pupa *(Randall Kaufmann)*

Hook: *100BL, 101 or 900BL, sizes 12-20*
Thread: *Color to match abdomen*
Tail: *White Antron, tied short*
Antennae: *Same piece as tail; short*
Rib: *White silk thread, twisted*
Abdomen: *Thread, dubbing, turkey biot, or pheasant tail feathers, thin*
Thorax: *Peacock*
Wing: *Wide grizzly hen hackle tips, tied one-third body length*
Note: For wing tie in, see page 70.

UFM

Comparadun *(Al Caucci and Bob Nastasi)*
Hook: 900BL, sizes 10-20
Thread: Color to match body
Wing: Mottled brown (or color to match natural) deer; single upright and flared
Tail: Blue dun hackle fibers, stiff, split wide
Abdomen: Color to match natural, Antron or Superfine

UFM

Conehead Woolly Sculpin, Tan *(Kelly Gallup)*
Hook: TMC 300, sizes 2-6
Head: Cone
Thread: Brown, 3/0
Tail: Cream marabou
Rib: Fine wire, color of choice
Abdomen: Brown/yellow variegated chenille, medium
Hackle: Cree or tan, palmered
Collar: Natural deer spun, clipped full
Head: Deer, trimmed to shape

UFM

Crystal Bugger, Maroon
Hook: 200RBL, 5262, 5263BL, sizes 6-10, weight to suit
Head: Bead, optional
Thread: Black
Tail: Maroon marabou and wine and red Krystal Flash
Rib: Medium red wire
Abdomen: Glo-Brite black Crystal Chenille, palmered with grizzly saddle dyed maroon
Notes: Try olive, black, red, purple, or brown.

UFM

Emergent Sparkle Pupa *(Gary LaFontaine)*
Hook: 900BL, sizes 12-20, weighted
Thread: Color of choice
Tail: Antron yarn to match veil
Abdomen Veil: Strands of Antron
Abdomen: Antron dubbing
Wing: Light deer hair, tied sparse
Head: Marabou or dubbing
Notes: Gary LaFontaine revolutionized fly fishing with Caddisflies, *and was one of fly fishing's most respected personalities.*

UFM

Ferry Canyon™ *(Randall Kaufmann, 1984)*
Hook: 7999, sizes 2-6, weighted
Thread: Black
Wing: Purple marabou over pearl, peacock, silver, red, blue, and dark purple Krystal Flash, or any other mix of reflectives.
Tail: Purple hackle fibers
Butt: Fl. Fire Orange Fuzzy Wool
Rib: Oval silver tinsel
Abdomen: Black chenille
Hackle: Purple

HF

Flash Fire Muddler *(Harold Lohr)*
Hook: TMC 5263BL, sizes 6-10
Head: Gold cone
Thread: White, 6/0; red, 6/0
Tail: Gold Flashabou dubbing
Abdomen: Gold Flashabou dubbing
Wing: White Z-Lon, gold Flashabou dubbing, and gold Krystal Flash
Collar: Deer
Head: Deer hair, trimmed to shape
Note: Also tied with pearl or silver.

UFM

Hare's Ear, Gold-Ribbed
Hook: 5262 or 3761BL, sizes 8-18, weighted
Thread: Brown
Tail: Hare's ear fur
Rib: Oval or flat gold mylar tinsel
Abdomen: Blended hare's ear
Wingcase: White tip turkey tail segment
Thorax: Blended hare's ear, picked out
Notes: Very popular nymph.

Hendrickson, Dark *(Roy Steenrod, 1916)*
Hook: 900BL, sizes 12-16
Thread: Gray or black
Wing: Mallard flank dyed woodduck
Tail: Dark blue dun hackle fibers
Abdomen: Dark gray, tan, or pinkish-tan dubbing
Hackle: Dark blue dun
Note: Old-time standard-style adult imitation of Ephemerella subvaria, *a major early-season hatch in the east. See page 68.*

Hot Butt® Emerger *(Randall Kaufmann)*
Hook: 3761BL, sizes 12-16
Thread: Black
Tail: Flame cerise or dark roe Glo-Bug yarn
Rib: Fire gold or copper wire
Abdomen: Peacock, pheasant tail, or any dubbing
Hackle: CDC tied 360°
Note: Drift, twitch, or tight line swing. Fish seem to come from nowhere—sometimes!

UFM

Hot Butt*® *Woolly Worm
Hook: *5263BL, sizes 8-10, weighted*
Thread: *Fluorescent fire orange*
Tail: *Glo-Bug yarn; pearl Krystal Flash*
Rib: *Fine wire, color of choice*
Abdomen: *White chenille, or color of choice*
Legs: *White saddle hackle or color of choice, palmered*
Note: This variation is from Joe Burke, an expert angler-tyer from Jackson Hole. Especially effective in the alpine lakes of the West. Try a grizzly and black combination.

Randy Stetzer

Lady Claret *(Randy Stetzer, 1985)*
Hook: *7999 or 7989 for low water style*
Thread: *Maroon*
Wing: *White calf tail, or kid goat*
Tail: *Claret hackle fibers*
Tag: *Oval silver tinsel*
Rib: *Flat or oval gold tinsel*
Abdomen: *Claret goat or Ice Dub*
Hackle: *Claret*
Note: Randy Stetzer is an accomplished tyer, northwest angling personality, and author of the best selling book, Flies: The Best 1,000.

UFM

Lead Eye Egg-Sucking Leech
Hook: *5263BL, sizes 2-6, weight to suit*
Thread: *Black*
Eyes: *Barbell—hourglass*
Tail: *Purple marabou with silver holographic Flashabou and purple and red Krystal Flash*
Rib: *Silver wire*
Abdomen: *Purple Crystal Chenille*
Hackle: *Purple saddle, palmered*
Head: *Fl. pink chenille*
Note: See page 62

UFM

Leech, Gold-Bead Peacock *(Brian Chan)*
Hook: *5263BL, sizes 8-10, weight to suit*
Head: *Gold bead*
Thread: *Black*
Tail: *Black marabou*
Abdomen: *Pearlescent black Crystal Chenille with pearl Flashabou*
Notes: Great colors, simple but excellent lake pattern. Brian Chan is an expert lake angler and tyer living in British Columbia.

UFM

Marauder™ *(Randall Kaufmann, 1995)*
Hook: *7999, sizes 2-6, weight to suit*
Thread: *Fluorescent fire orange*
Abdomen: *Pearlescent Estaz or Crystal Chenille*
Wing: *Mirror Flash, Krystal Flash, and Flashabou—pinks, reds, pearl, gold, and holographic*
Hackle: *Cerise, soft and long*
Note: Could be called the Silver King. Popular silver salmon pattern. Customize colors.

HF

Matuka
Hook: *5263BL or 300, sizes 4-8, weight to suit*
Thread: *To match abdomen*
Rib: *Silver, or gold wire, or oval tinsel*
Abdomen: *Chenille*
Throat: *Red SLF or other dubbing*
Wing: *Hen or soft and wide saddle hackles*
Hackle: *Hen or soft saddle*
Eyes: *Paint to suit*

UFM

***Parachute,* Baetis** *(Pret Frazier)*
Hook: *900BL, sizes 16-22*
Thread: *Olive*
Wing: *Mallard*
Tail: *Blue dun Micro Fibetts, two, split*
Abdomen: *Olive to olive-brown Superfine*
Hackle: *Dyed olive grizzly or natural grizzly*
Note: One of the better Baetis *patterns. See page 74.*

UFM

Parachute, Black Gnat
Hook: *900BL, sizes 14-18*
Thread: *Black*
Wing: *White calf, parachute style*
Tail: *Black hackle fibers*
Abdomen: *Black dubbing*
Hackle: *Black, parachute style*
Note: The black and white color combination offers anglers visibility under most light conditions. See page 74.

UFM

Parachute, Light Cahill
Hook: *900BL, sizes 12-18*
Thread: *Pale yellow or cream*
Wing: *White calf, parachute style*
Tail: *Light ginger or cream hackle fibers*
Abdomen: *Cream dubbing*
Hackle: *Light ginger or cream, parachute style*
Note: See page 74.

UFM

Pheasant Tail, Theo's *(Theo Bakelaar)*
Hook: 3761BL, sizes 12-16, weight to suit
Head: Gold bead
Thread: Tan
Tail: Brown ringneck pheasant tail
Rib: Fine copper wire
Abdomen: Brown ringneck pheasant tail
Thorax: Tan Antron
Notes: Theo is an inventive angler/tyer living in Holland. He was one of the first to promote bead head flies in America.

UFM

Polar Shrimp *(1930s)*
Hook: 7999, sizes 2-6
Thread: Black
Wing: White calf or bucktail
Tail: Hot or fluorescent orange hackle fibers
Abdomen: Fluorescent orange chenille
Hackle: Hot or fluorescent orange hackle
Note: All around Northwest, Canadian and Alaska favorite—steelhead, rainbow, Dollies, char, and Pacific salmon.

UFM

Prince *(Doug Prince, 1950s)*
Hook: 5262, 3761BL, or 200RBL, sizes 4-10, weighted
Thread: Black
Tail: Brown goose or turkey biot, tied in V
Rib: Flat gold tinsel
Abdomen: Peacock
Legs: Brown hackle, 3 to 5 turns, tied full
Wing: White goose or turkey biot, tied flat over body in V *with tips curving up.*

UFM

Purple Peril *(Ken McLeod, 1940s)*
Hook: 7999, sizes 2-6
Thread: Black
Wing: Red fox squirrel or bucktail
Tail: Purple hackle fibers
Rib: Flat silver tinsel
Abdomen: Purple chenille
Hackle: Purple
Notes: One of the few "old time" survivors, and forerunner of current purple patterns

Quill Gordon *(Theodore Gordon)*
Hook: 900BL, sizes 12-18
Thread: Gray
Wing: Mallard dyed woodduck, upright and divided
Tail: Blue dun hackle fibers
Abdomen: Stripped peacock quill
Hackle: Blue dun
Note: One of the longest-surviving fly patterns. Represents Epeorus pleuralis. *See March Brown, page 68.*

UFM

Renegade *(Taylor "Beartracks" Williams)*
Hook: 900BL, sizes 10-16
Thread: Black
Tag: Flat gold tinsel
Aft Hackle: Brown
Body: Peacock
Fore Hackle: White
Note: This is one of the few flies that has fore and aft hackle. "Beartracks" Williams first tied the Renegade in 1928. George L. Herter claimed the fly to be an English pattern tied by Mrs. Edith Cox in the late 1800s.

UFM

San Juan Worm
Hook: 2457, sizes 12-16
Thread: To match body
Abdomen: Ultra Chenille (red, maroon, orange, brown, or tan) and Ultra Wire
Notes: Represents garden hackle.

Simulator *(Randall Kaufmann, 1970s)*
Hook: 5263BL or 200RBL, sizes 6-12, weighted
Thread: Olive
Tail: Olive marabou
Rib: Copper wire
Abdomen: Peacock
Hackle: Saddle hackle, palmered, clipped
Notes: Versatile. Vary the color, size, and shape—cranefly, dobsonfly (hellgrammite), *beetle larvae, stones, and caddis.*

Timberline Emerger *(Randall Kaufmann)*
Hook: 3761BL, sizes 12-16
Thread: Gray
Tail: Natural gray fluff from ringneck body feather, short and heavy
Rib: Copper wire
Abdomen: Gray SLF or other sparkly dubbing
Wing: Grizzly hen hackle tips
Legs: Brown hackle or CDC tied full
Notes: Try other colors, especially olive or tan.

Index